GRAPH THEORY AND FEYNMAN INTEGRALS

Mathematics and Its Applications

A Series of Monographs and Texts Edited by
Jacob T. Schwartz, Courant Institute of Mathematical Sciences,
New York University

Volume 1

Jacob T. Schwartz LECTURES ON THE MATHEMATICAL METHOD IN ANALYTICAL ECONOMICS

Volume 2

O. A. Ladyzhenskaya THE MATHEMATICAL THEORY OF VISCOUS INCOMPRESSIBLE FLOW (*Second English Edition. Revised and Enlarged*)

Volume 3

R. L. Stratonovich TOPICS IN THE THEORY OF RANDOM NOISE (*in 2 volumes*)

Volume 4

S. Chowla THE RIEMANN HYPOTHESIS AND HILBERT'S TENTH PROBLEM

Volume 5

Jacob T. Schwartz THEORY OF MONEY

Volume 6

F. Trèves LINEAR PARTIAL DIFFERENTIAL EQUATIONS WITH CONSTANT COEFFICIENTS

Volume 7

Georgi E. Shilov GENERALIZED FUNCTIONS AND PARTIAL DIFFERENTIAL EQUATIONS

Volume 8

I. I. Pyatetskii-Shapiro AUTOMORPHIC FUNCTIONS AND THE GEOMETRY OF CLASSICAL DOMAINS

Volume 9

I. Colojoară and C. Fioaș THEORY OF GENERALIZED SPECTRAL OPERATORS

Volume 10

M. N. Barber and B. W. Ninham RANDOM AND RESTRICTED WALKS

Volume 11

N. Nakanishi GRAPH THEORY AND FEYNMAN INTEGRALS

Additional volumes in preparation

Graph Theory and Feynman Integrals

NOBORU NAKANISHI

Kyoto University, Japan, and
Brookhaven National Laboratory, New York

GORDON AND BREACH
Science Publishers
NEW YORK LONDON PARIS

Editorial office for the United Kingdom
Gordon and Breach, Science Publishers Ltd.
12 Bloomsbury Way
London W.C. 1

Editorial office for France
Gordon & Breach
7-9 rue Emile Dubois
Paris 14^{e}

Library of Congress catalog card number 71-123483. AMS 1970 Subject Classifications 81 A 15, 05 C 99, 81 A 51. ISBN 0 677 02950 0. Printed in east Germany.

Dedicated to

Dr. Yoshio Shimamoto

who was the first to combine graph theory with Feynman integrals

Preface

The purpose of this book is to present a rather detailed survey of the graph-theoretical aspects of Feynman integrals. The reader is supposed to be *either* a theoretical physicist who is familiar with Feynman graphs and Feynman rules *or* a graph theoretician who may not be interested in the elementary-particle physics. This book, therefore, contains neither the derivation of Feynman integrals from quantum field theory nor their practical computations in the scattering problems. Excellent presentations on those subjects can be found in many standard textbooks of quantum field theory. The problems studied in this book are graph-theoretical and analytic properties of the *general-order* Feynman integral written in the Feynman-parametric form. The homological approach to the Feynman integral is not discussed.

Each chapter contains its own introduction, in which a brief historical review of the research is sketched. Proofs of theorems and other propositions given in this book are not necessarily the ones given in the original papers, but a number of new improved proofs are presented; also included are some new results which become necessary for the systematic description.

In Chapter 1, graph theory is reviewed as the preliminaries for the subsequent chapters. I would apologize to graph theoreticians that the graph-theoretical notation adopted here is not the one used in most literatures of graph theory, though the latter is not yet well established. Chapter 2 is the central part of this book. After explaining briefly physical background of the Feynman integral, the general Feynman-parametric formula is derived from the Feynman integral, and the properties of the integrand of the former are investigated in detail. The Feynman-parametric formula is applied to the study of the analytic properties of the Feynman integral in the subsequent two chapters. Chapter 3 deals with the singularities of *each* Feynman integral, while Chapter 4 concerns analyticity domains which are common to a *class* of Feynman integrals. Some miscellaneous topics are discussed in Chapter 5. Appendix A is added for the convenience of the reader who is not familiar with distributions and analytic functions. In

Appendix B, important results concerning the single-loop Feynman graphs are summarized without proof.

Throughout this book, concepts and properties on matrices and determinants are used without explanation; for example, the reader is assumed to know a cofactor, a minor, Jacobi's theorem, etc. Some standard symbols are used without notice; for example, det, sup, and inf mean "the determinant of", "the supremum of", and "the infimum of", respectively.

References are given at the end of each chapter. They may not be complete; especially, many papers on the Feynman integrals are omitted if they do not have graph-theoretical aspects. The year stated as that in which the work was done should be understood as the "received" date, that is, the year when the corresponding paper was submitted for publication.

I would like to express my sincere gratitude to Dr. Y. Shimamoto, who led me to the graph-theoretical consideration of the theory of Feynman integrals several years ago and suggested I write a book on this subject. I am very grateful to several friends of mine at Kyoto University, especially to Professor H. Araki and to Dr. M. Minami, for useful discussions in the preliminary stage of the manuscript. I am also very much indebted to Dr. E. P. Speer for critically reading a preliminary manuscript and making many interesting comments. Thanks are also due to Misses K. Hatsuda, M. Inoue, T. Maeda, and R. Miki for typing the preliminary manuscripts and to Mrs. N. Griffin for typing the final manuscript. A part of the final manuscript was written while I was a Visiting Physicist in the Applied Mathematics Department of Brookhaven National Laboratory. Finally, I would like to thank Dr. J. Weneser, Professor J. Schwartz, Dr. E. H. Immergut, and Mrs. E. A. Pusey for making it possible to have this book published by Gordon and Breach, Science Publishers, Inc.

N. Nakanishi

Contents

CHAPTER 1

Graph Theory

The objects of graph theory are linear graphs. A linear graph consists of a number of lines (or edges) and their end points, called vertices. Historically, graph-theoretical considerations were developed first as amusements of mathematicians such as the Königsberg bridge problem, Hamilton's round tour of the world, the four-color problem of planar maps, etc. Graph-theoretical objects were encountered also in various branches of not only natural sciences but also cultural sciences. Among others, the theory of electrical networks, which was originated by Kirchhoff in the middle of the 19th century, made an important contribution to the algebraic studies of graphs. The establishment of graph theory as a branch of mathematics is, however, rather new (maybe not earlier than 1930), and because of this historical situation the terminology of graph theory is quite disorganized. Nowadays, a number of books concerning graph theory are available [1, 2, 3, 4], but different authors often use different words for the same concept.

The Feynman graph, which was introduced into quantum field theory first by Feynman in 1949, presents a new and interesting application of graph theory. The structure of a Feynman graph is so akin to that of an electrical network that many results obtained in the latter are very useful also in the former. The main purpose of the present chapter is to describe those algebraic studies of graphs. We do not intend to present a complete survey of graph theory.

In Section 1, we explain terminology and notation, which will be used throughout this book. Important graph-theoretical concepts are paths, circuits, cut-sets, and trees. Roughly speaking, they are the following objects: A path is a set of lines linking two veritces; a circuit is a set of lines which makes topologically a circle; a cut-set is a set of lines dividing a connected graph into exactly two parts; a tree is a set of lines which includes no circuit but links all vertices. In Section 2, some fundamental properties of those sets of lines are discussed. We emphasize the importance of the duality between circuits and cut-sets. In Section 3, we introduce the incident, the circuit, and the cut-set matrix, and derive various topological formulas for the determinants of products of matrices. They will become important in the consideration of the Feynman-parametric integral (see Chapter 2). The remaining two sections are devoted to some special topics. In Section 4, we present a sketch of planar graphs and dual graphs as preliminaries of the dual-graph analysis of the Landau equations (see Subsections 12-2 and 13-2). In Section 5, the minimum cut theorem and its generalization are described almost solely because they will be applied later in order to find support properties of certain integral representations (see Section 19).

§ 1 TERMINOLOGY AND NOTATION

1-1 Set-Theoretical Concepts

We shall extensively employ the notation of set theory. A *set* is a collection of definite things, called *elements*. The set consisting of elements α, β, γ, ... is denoted by $\{\alpha, \beta, \gamma, \ldots\}$, and the set of all elements α satisfying certain conditions $\Gamma(\alpha)$ is expressed as $\{\alpha \mid \Gamma(\alpha)\}$. If α is an element of a set A, we write $\alpha \in A$ or $A \ni \alpha$. A set B is a *subset* of A if all elements of B belong to A. If B is a subset of A, we write $B \subset A$ or $A \supset B$. For two sets A and B, we have $A = B$ if and only if $A \supset B$ and $A \subset B$. The symbols $\notin$, $\not\ni$, $\not\subset$, $\not\supset$, $\neq$ denote the negatives of $\in$, $\ni$, $\subset$, $\supset$, $=$, respectively. A subset B of A is called *proper* if $B \neq A$. The set consisting of no elements is called the *empty set* and denoted by $\emptyset$.

Given two sets A and B, a *union* of A and B is the set consisting of all elements which belong to either A or B, and an *intersection* of A and B is the set consisting of all elements which belong to both A and B. We denote a union of A and B by $A \cup B$ and an intersection of them by $A \cap B$. A union and an intersection of more than two sets can similarly be defined, and a union of $A_1, A_2, \ldots, A_k$ is denoted by $\bigcup_{j=1}^{k} A_j$. For $\alpha \notin A$, we denote $A \cup \{\alpha\}$ by $A + \alpha$. If $A \cap B = \emptyset$, then A and B are called *disjoint*. When any two of $A_1, \ldots, A_k$ are disjoint, $\bigcup_{j=1}^{k} A_j$ is called a *disjoint union*. We denote by $A - B$ the set which consists of all elements of A not contained in B. If $B \subset A$, then $A - B$ is called the *complement* of B and sometimes denoted by B^* if there is no possibility of confusion. For $\alpha \in A$, $A - \{\alpha\}$ is abbreviated as $A - \alpha$.

The *symmetric difference* of two subsets B_1 and B_2 of a set A is defined by $B_1 \cup B_2 - B_1 \cap B_2$ and denoted by $B_1 \oplus B_2$. The operation $\oplus$ is commutative and associative.

For two sets A and B, a *mapping* from A to B is a correspondence such that any element of A corresponds to only one element of B. Let φ be a mapping from A to B. If $\alpha \in A$ corresponds to $\beta \in B$ by φ, we write $\beta = \varphi(\alpha)$. For $C \subset A$, we define $\varphi(C)$ by $\{\varphi(\alpha) \mid \alpha \in C\}$. Then $\varphi(A)$ is the *image* of A and of course a subset of B. If $B = \varphi(A)$ then φ is called *onto*. If φ is onto and if there exists a mapping φ^{-1} from $B = \varphi(A)$ to A such that $\varphi^{-1}(\varphi(\alpha)) = \alpha$ for any $\alpha \in A$, then A and B are in *one-to-one correspondence*. For $C \subset A$, the *restriction* of φ to C is a mapping φ' from C to B such that $\varphi'(\alpha) = \varphi(\alpha)$ for any $\alpha \in C$.

A relation (denoted by $\simeq$) defined in a set A is an *equivalence relation* if $\alpha \simeq \alpha$, if $\alpha \simeq \beta$ implies $\beta \simeq \alpha$, and if $\alpha \simeq \beta$ and $\beta \simeq \gamma$ imply $\alpha \simeq \gamma$ for any $\alpha, \beta, \gamma \in A$. Then A is a disjoint union of subsets B_j such that $\alpha \simeq \beta$ for any $\alpha, \beta \in B_j$ but not for any $\alpha \in B_j$ and $\beta \in B_k$ $(k \neq j)$. The sets B_j are called *equivalence classes.*

A set A having a property Γ is called *minimal* if any proper subset of A does not have the property Γ. Likewise, a subset B of a set A is called *maximal* if any subset of A containing B as its proper subset does not have the property Γ.

An *additive function* $f(B)$ over a set A is a function of $B \subset A$ such that $f(B_1 \cup B_2) = f(B_1) + f(B_2)$ for any disjoint subsets B_1 and B_2.

1-2 Graphs

A *graph* (more precisely, a *linear graph*) consists of two kinds of elements, called *lines* and *vertices*, together with an incidence relation between them. An *incidence relation* is a mapping from the set of all lines to the set of pairs of vertices (the two vertices in each pair may not necessarily be distinct). For a graph G, the set of all lines of G is denoted by $|G|$ and that of all vertices of G is denoted by $v(G)$. Let φ be the incidence relation of G. For $l \in |G|$, we can write

$$\varphi(l) = (a, b) \quad \text{for } a, b \in v(G) \tag{1-1}$$

by definition. Then a and b are called the *end points* of a line l, and l is called *incident* with a (and b). In (1-1), if (a, b) and (b, a) are not distinguished, G is called an *unoriented graph*. Otherwise G is an *oriented graph*.† We shall always consider an oriented graph. In this case, in (1-1), a is called the *outgoing* end point and b is called the *ingoing* end point.

Intuitively, a graph can be realized geometrically in a three-dimensional euclidean space. A vertex and a line are represented as a point and an arc (with an orientation), respectively. The incidence relation (1-1) is represented by linking two points a and b by an arc l. We also note that a graph is a concept which is essentially the same as that of a one-dimensional complex in combinatorial topology if two end points of any line are distinct.

† An oriented graph is different from a *directed* graph, which has intrinsic orientations. In the former, one can change orientations without bringing any essential change of results.

In principle, the numbers of lines and vertices may be infinite, but we always consider only the case in which they are both finite. Such a graph is called a *finite graph*; in the following a graph always means a finite graph.

A *subgraph*, H, of a graph G is a graph such that $|H| \subset |G|$ and $v(H) \subset v(G)$ and the incidence relation of H is the restriction of φ to $|H|$. For a graph G, $N(G)$ and $M(G)$ denote the number of lines of G and that of vertices of G, respectively. For a subset I of $|G|$, $N(I)$ stands for the number of lines of I. It is an additive function of I over $|G|$.

A *star* is the set of all lines incident with a particular vertex, say a, and it is denoted by $S[a]$. If $S[a] = \emptyset$, a is called an *isolated point*. A line whose end points are not distinct is called a *loop line*, that is, l is a loop line if $\varphi(l) = (a, a)$. The set of all loop lines incident with a is denoted by $L[a]$. Of course $L[a] \subset S[a]$. The *degree* of a vertex a is defined to be

$$N(S[a]) + N(L[a]) \tag{1-2}$$

that is, the number of times by which a appears in the right-hand side of (1-1) when l runs over $|G|$. The *relative degree* with respect to a subset I of $|G|$ is defined to be

$$D(a, I) \equiv N(S[a] \cap I) + N(L[a] \cap I) \tag{1-3}$$

We see that $D(a, I)$ is an additive function of I, whence

$$\sum_{a \in v(G)} D(a, I) = 2N(I). \tag{1-4}$$

Two or more lines are called *parallel* if they belong to $S[a] \cap S[b]$ for some a and b ($a \neq b$). For a set I, if $I \cap S[a] \neq \emptyset$ then we say that I *passes through* the vertex a. If I passes through a, any subgraph H such that $|H|$ equals I has to contain a.

Two distinct vertices a and b are called *adjacent* if $S[a] \cap S[b] \neq \emptyset$. Two vertices a and b, which may not necessarily be distinct, are called *connected* if there is a sequence of vertices $a = a_0, a_1, a_2, \ldots, a_k = b$ such that a_j and a_{j+1} ($j = 0, 1, \ldots, k - 1$) are adjacent. The concept "connected" is an equivalence relation, whence $v(G)$ is decomposed into equivalence classes. By definition, no line can be incident with two vertices which belong to different equivalence classes. Thus $|G|$ also is decomposed into disjoint subsets each of which is associated with an equivalence class of vertices. A *connected component* is a subgraph of G defined by an equivalence class of vertices and the corresponding set of lines. The number of connected components of G is denoted by $p(G)$. A graph G is a *connected graph* if it consists of only one connected component (i.e., $p(G) = 1$). A set of lines I is called *connected* if there is a connected subgraph H such that $|H| = I$.

To *delete* a line l of G is to consider a subgraph of G such that the set of its lines is $|G| - l$ and that of its vertices is $v(G)$. We denote this subgraph by $G - l$. For a set of lines I, we denote by $G - I$ the graph which is obtained by deleting all lines belonging to I. A *cut-line* l is a line such that the number of connected components of $G - l$ is greater than that of G (by one). A *strongly connected* (or *cyclically connected*) graph is a connected graph which contains no cut-lines.

A vertex a is a *cut-vertex* (or an *articulation point*) if a connected component H of G has two subgraphs H_1 and H_2 such that $|H_1| \neq \emptyset$, $|H_2| \neq \emptyset$, $|H_1| \cap |H_2| = \emptyset$, $|H_1| \cup |H_2| = |H|$, and $v(H_1) \cap v(H_2) = \{a\}$. Intuitively, a cut-vertex is a point such that a connected component of G becomes geometrically disconnected when its "neighborhood" is removed. A *nonseparable* graph is a connected graph which contains no cut-vertex. A nonseparable graph G is strongly connected except for the case in which $|G|$ consists of only one cut-line, because an end point of a cut-line l is a cut-vertex if a line other than l is incident with it.

To *identify* two vertices a and b of G is to consider a graph consisting of $|G|$ and $v(G) - b$ together with an incidence relation φ', where φ' is defined from φ by replacing b by a (in this description, we may interchange a and b). This graph is denoted by $G^{(ab)}$ ($\equiv G^{(ba)}$) (see Fig. 1-1). To *contract* a line l of G is to delete l and then identify the two end points of l. Intuitively, to contract l is to shrink it to a point. We denote by G/l the graph which is

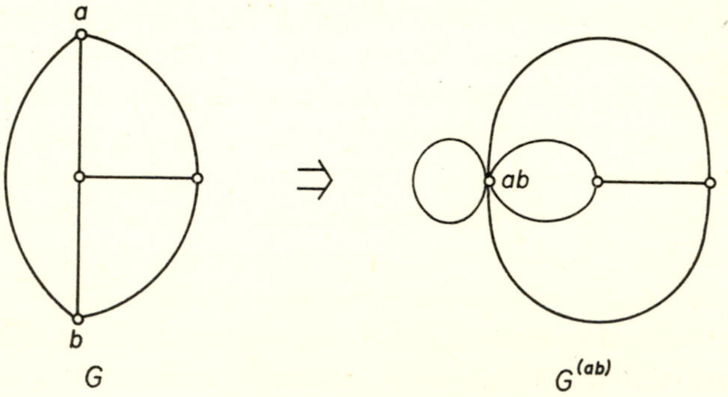

Fig. 1-1 An example of $G^{(ab)}$. It is obtained from G by identifying two vertices a and b.

obtained from G by contracting a line l; likewise G/I stands for the graph obtained by contracting all lines of I. We call G/I a *reduced graph* of G.

A *complete graph* is a graph such that for any two distinct vertices a and b we have $N(S[a] \cap S[b]) = 1$ and also $L[a] = \emptyset$ for any $a \in v(G)$. However, we also call a graph such that $S[a] \cap S[b] \neq \emptyset$ for any $a \neq b$ a complete graph in the generalized sense (see Section 19). The complete graphs having 2, 3, 4, 5 vertices are shown in Fig. 1-2.

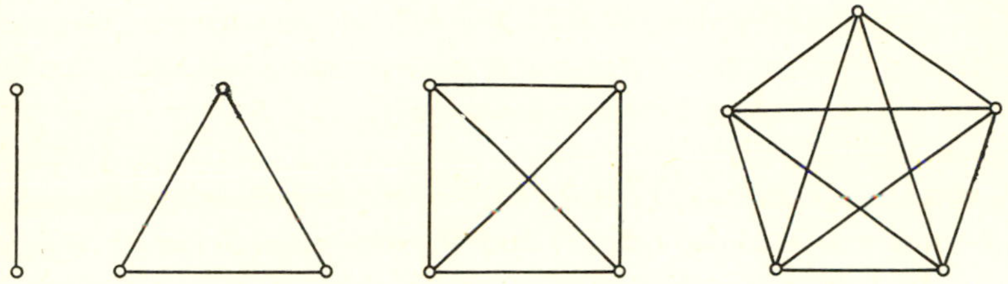

Fig. 1-2 Complete graphs having two, three, four, and five vertices

A *planar graph* is a graph which is geometrically realizable on a plane. See Section 4 for details.

1-3 Paths, Circuits, Cut-Sets, Trees

In this subsection, we define some important concepts such as paths, circuits, cut-sets, trees, etc. In the literatures, they are regarded as subgraphs, but here, for simplicity of notation, they are defined to be sets of lines. In the following, we consider some sets of lines in an arbitrary graph G.

A *path* P between two distinct vertices a and b is a minimal connected set of lines such that P passes through both a and b. By definition, P exists if and only if a and b are connected. Hence there exists a sequence of vertices $a = a_0, a_1, \ldots, a_k = b$ such that $S[a_j] \cap S[a_{j+1}] \cap P \neq \emptyset$ for $j = 0, 1, \ldots, k - 1$. Since P is minimal, therefore, it can also be defined as follows. A path P between two distinct vertices a and b is a connected set of lines such that $D(a, P) = D(b, P) = 1$ and $D(c, P) = 0$ or 2 for any other vertex c. The vertices a and b are called the *end points* of P.

The set of all paths between a and b in a graph G is denoted by $\boldsymbol{P}_G(ab)$ or simply by $\boldsymbol{P}(ab)$ $(= \boldsymbol{P}(ba))$. It is convenient to define $\boldsymbol{P}(aa)$ to be a set consisting of the empty set alone. The statement that a and b are connected is equivalent to $\boldsymbol{P}(ab) \neq \emptyset$.

A *circuit* (or a *loop*) C is a minimal nonempty set of lines such that $D(a, C) \neq 1$ for any vertex a. By definition, we can find a sequence l_1, $l_2, \ldots, l_k$ of *distinct* lines belonging to C such that l_j and l_{j+1} are incident with a vertex a_j ($j = 1, 2, \ldots, k - 1$), ($k \geqq 1$). Since k cannot be arbitrarily large, we have to have a maximal one. Then the other end point a_k of l_k should be identical with one of $a_0, a_1, \ldots, a_{k-1}$ because $D(a_k, C) \geqq 2$, where a_0 is the other end point of l_1. Since C is minimal, therefore, it can also be defined as follows. A circuit C is a nonempty connected set of lines such that $D(a, C) = 0$ or 2 for any vertex a.

The set of all circuits in a graph G is denoted by $\boldsymbol{C}_G$ or simply by $\boldsymbol{C}$.

A *cut-set* S is a minimal set of lines such that a connected component of G becomes disconnected in $G - S$. In a connected graph G, a cut-set S divides $v(G)$ into two disjoint subsets u and u^*, the equivalence classes in $G - S$.

The set of all cut-sets in a graph G is denoted by $\boldsymbol{S}_G$ or simply by $\boldsymbol{S}$. In a connected graph, we denote by $\boldsymbol{S}(u \mid u^*)$ the set of all cut-sets which divide $v(G)$ into u and u^*. Let h and h' be two arbitrary disjoint subsets of $v(G)$; then $\boldsymbol{S}(h \mid h')$ $(= \boldsymbol{S}(h' \mid h))$ denotes the union of $\boldsymbol{S}(u \mid u^*)$ over all u such that $u \supset h$ and $u^* \supset h'$. If $S \in \boldsymbol{S}(h \mid h')$ then for any $a \in h$ and $b \in h'$ a and b are disconnected in $G - S$. When $h = \{a, \ldots, b\}$ and $h' = \{c, \ldots, d\}$, we write $\boldsymbol{S}(h \mid h')$ as $\boldsymbol{S}(a, \ldots, b \mid c, \ldots, d)$; in particular, $\boldsymbol{S}(a \mid b) = \boldsymbol{S}(\{a\} \mid \{b\})$.

Let u and u^* be two disjoint subsets of $v(G)$ such that $u \cup u^* = v(G)$. A *cut* is the set of all lines which are incident with both a vertex of u and that of u^*, that is, a cut "divides" $v(G)$ into u and u^*. A cut-set is a cut, but the converse is not true.

A *forest* is a maximal set of lines which includes no circuit. If G is connected, a forest is called a *tree*. If T is a tree and if $l \in T$ and $l' \notin T$, then $T^* = |G| - T$ is a *chord set*, $T - l$ is a 2-*tree*, and $T + l'$ is a *pseudotree*. Furthermore, if $I \subset T$ and if $\mathrm{N}(I) = n - 1$ then $T - I$ is an *n-tree*.

The set of all trees and that of all chord sets are denoted by $\boldsymbol{T}_G$ or $\boldsymbol{T}$ and by $\boldsymbol{T}^*_G$ or $\boldsymbol{T}^*$, respectively. They are always nonempty. Some sets of 2-trees or n-trees will be defined in Section 2.

1-4 Feynman Graphs

A *Feynman graph* is a graph each line of which topologically represents a propagation of a free elementary particle and each vertex of which represents an interaction of elementary particles. For more precise explanation, see Section 6. Here, we regard a Feynman graph as an abstract object.

A Feynman graph is a graph which may have some extraordinary lines, called *external lines*, in addition to the ordinary lines, which are sometimes called *internal lines*. Every external line has only one end point, that is, if j is an external line, $\varphi(j)$ is a vertex but not a pair of vertices (for simplicity, we exclude the case in which some external lines are not incident with any vertex at all). A vertex is called an *external vertex* if at least one external line is incident with it. Vertices other than external vertices are called *internal vertices*. The set of all external vertices is usually denoted by g.

All concepts defined in Subsections 1-2 and 1-3 can be used also in a Feynman graph G. It is convenient, however, to introduce some additional concepts. A *sub-Feynman-graph* H of G is a Feynman graph such that H is a subgraph of G in the sense of a graph and the external lines of H are the external and internal lines of G whose only one end point belongs to $v(H)$. The *F-degree* of a vertex a is a sum of the degree of a and the number of the external lines incident with a. Let h be a subset of g such that $h \neq \emptyset$ and $g - h \neq \emptyset$. Then a division $(h \mid g - h)$ of g is called a *channel*. If a cut-set S belongs to $\boldsymbol{S}(h \mid g - h)$, then it is called an *intermediate state* in the channel $(h \mid g - h)$. Planar Feynman graphs will be defined in Section 4. A *complete Feynman graph* is a Feynman graph such that it is complete as a graph and all its vertices are external.

According to the total number n of external lines, connected Feynman graphs have various names. For $n = 0$, $n = 1$, $n = 2$, $n = 3$, $n = 4$, and $n = 5$, they are called *vacuum polarization graphs*, *tadpole graphs*, *self-energy graphs*, *vertex graphs*, *two-particle scattering graphs*, and *one-particle production graphs*, respectively. Likewise, connected sub-Feynman-graphs are called *vacuum polarization parts*, *tadpole parts*, *self-energy parts*, *vertex parts*, etc.

We sometimes call a Feynman graph simply a graph if there is no possibility of confusion.

§2 FUNDAMENTAL PROPERTIES OF A GRAPH

Throughout this and next sections, we suppose that a graph G has N lines and M vertices, namely, $N(G) = N$ and $M(G) = M$.

2-1 Properties Independent of Orientation

The following properties are direct consequences of the definitions given in Section 1.

Any loop line constitutes a circuit by itself. It intersects no cut-set, no cut, and no forest. Loop lines are only lines which are not contained in any cut-set. A nonseparable graph G cannot contain a loop line unless $|G|$ consists of this loop line alone.

Any cut-line constitutes a cut-set by itself. It intersects no circuit. Cut-lines are only lines which are not contained in any circuit. Any forest contains all cut-lines.

If I is a nonempty set of lines such that $D(a, I) \neq 1$ for any vertex a, then I includes a circuit by definition. Hence, for any forest T, there exists at least one vertex a such that $D(a, T) = 1$.

If I is a set of lines such that $D(a, I)$ is even for any vertex a, then I is a disjoint union of circuits, because if C is a circuit included in I then $I - C$ again satisfies the assumption. If $P_1, P_2 \in \boldsymbol{P}(ab)$ then $P_1 \oplus P_2$ is a disjoint union of circuits. In particular, if $P_1 \neq P_2$ then $P_1 \cup P_2$ includes a circuit. If $C_1, C_2 \in \boldsymbol{C}$ then $C_1 \oplus C_2$ is a disjoint union of circuits.

Any graph G is uniquely decomposed into maximal nonseparable subgraphs $G_1, G_2, \ldots, G_k$ in such a way that $|G_j| \neq \emptyset$ for any j, $|G_i| \cap |G_j| = \emptyset$ for any $i \neq j$, and $\bigcup_{j=1}^{k} |G_j| = |G|$. Any $C \in \boldsymbol{C}_G$ (or any $S \in \boldsymbol{S}_G$) is included in only one of $G_1, \ldots, G_k$.

THEOREM **2-1** Given a graph G, let $I \subset |G|$.

(a) If $G - I$ is connected [or strongly connected], then G is connected [or strongly connected].

(b) If $I \cap P = \emptyset$ then $P \in \boldsymbol{P}_G(ab)$ implies $P \in \boldsymbol{P}_{G-I}(ab)$ and *vice versa*.

(c) If $I \cap C = \emptyset$ then $C \in \boldsymbol{C}_G$ implies $C \in \boldsymbol{C}_{G-I}$ and *vice versa*.

(d) If $S \not\subset I$ and if $S \in \boldsymbol{S}_G$ then there exists $S' \in \boldsymbol{S}_{G-I}$ such that $S' \subset S - I$.

Proof (a). Evident. (b) and (c). We note that $D(a, P)$ and $D(a, C)$ for any vertex a are independent of I. (d). The number of the connected components of $G - (I \cup S)$ is greater than that of $G - I$. Hence $S - I$ includes a cut-set in $G - I$. q.e.d.

THEOREM **2-2** Given a graph G, let $I \subset |G|$.

(a) If G is connected [or strongly connected], then G/I is connected [or strongly connected].†

(b) If $P \in \boldsymbol{P}_G(ab)$ then there exists $P' \in \boldsymbol{P}_{G/I}(ab)$ such that $P' \subset P - I$.

(c) If $C \not\subset I$ and if $C \in \boldsymbol{C}_G$ then there exists $C' \in \boldsymbol{C}_{G/I}$ such that $C' \subset C - I$.

(d) If $I \cap S = \emptyset$ then $S \in \boldsymbol{S}_G$ implies $S \in \boldsymbol{S}_{G/I}$ and *vice versa*.

Proof (a) and (b). Evident. (c). Relative degrees of all vertices with respect to $C - I$ are even in G/I. (d). The defining properties of S are independent of the contraction of I. q.e.d.

THEOREM **2-3** Let a be any vertex of a graph G.

(a) If G is nonseparable and $M \geqq 2$, then $S[a]$ is a cut-set.
(b) In general, $S[a] - L[a]$ is a disjoint union of cut-sets.

Proof (a). By definition, $S[a]$ has to include a cut-set S if G is connected and $M \geqq 2$, because then a is disconnected from any other vertex in $G - S[a]$. If $S[a] - S \neq \emptyset$ then a is a cut-vertex. (b). In $G/L[a]$, $S[a] - L[a]$ is either empty or a set including a cut-set, say S_1. For the latter case, in $G/(L[a] \cup S_1)$, $S[a] - (L[a] \cup S_1)$ is again so. Thus we can proceed by induction. We note that S_1, etc. are cut-sets in G because of Theorem 2-2 (d). q.e.d.

THEOREM **2-4** Let G be a connected graph.

(a) A cut $\bar{S}$ is a cut-set if and only if $G - \bar{S}$ consists of two connected components.

(b) In general, a cut $\bar{S}$ is a disjoint union of cut-sets.

Proof (a). By definition, $\bar{S}$ includes a cut-set. Since any line of $\bar{S}$ is incident with vertices of both connected components, $\bar{S}$ is minimal. (b). Suppose that any line of $\bar{S}$ is incident with a vertex of u and that of $u^* = v(G) - u$. Then, in $G/\bar{S}^*(\bar{S}^* = |G| - \bar{S})$, any two adjacent vertices cannot both belong to either u or u^* alone. Therefore $\bar{S}$ is a disjoint union of stars in $G/\bar{S}^*$. Since $\bar{S} \cap L[a] = \emptyset$, each star is a disjoint union of cut-sets

† Note that if G is nonseparable G/I is not necessarily nonseparable.

in $G/\bar{S}^*$ because of Theorem 2-3 (b). Thus $\bar{S}$ is a disjoint union of cut-sets in G because of Theorem 2-2 (d). q.e.d.

THEOREM **2-5** A tree T of a connected graph G has the following properties:

(a) T passes through all vertices of G (if $M \geqq 2$).

(b) T is connected.

(c) For any $a, b \in v(G)$ $(a \neq b)$, there is a unique path between a and b included in T.

(d) $N(T) = M - 1$ [hence $N(T^*) = N - M + 1$].

Proof (a). If $S[a] \cap T = \emptyset$ then for $l \in S[a] - L[a]$, $T + l$ includes no circuit. This contradicts the maximality of a tree. (b). If T is disconnected, then T^* includes a cut-set S. Then for $l \in S$, $T + l$ includes no circuit. (*c*). The existence of a path follows from (a) and (b). It is unique because T includes no circuit. (d). Since it is trivially true for $M = 1$, we employ induction. Since there is a vertex, say a, whose relative degree with respect to T is 1, we have $N(T - S[a]) = N(T) - 1$. In G/l with $\{l\} = T \cap S[a]$, $T - S[a]$ is a tree because it includes no circuit and it is maximal [see Theorem 2-2 (c)]. Hence $N(T - S[a]) = M(G/l) - 1 = M - 2$ by induction assumption. Thus $N(T) = M - 1$. q.e.d.

THEOREM **2-6** In a graph G, any of the following systems of properties implies that T is a tree.

(a) T includes no circuit, T is connected, and T passes through all vertices of G.

(b) T includes no circuit and $N(T) \geqq M - 1$.

(c) T is connected, T passes through all vertices of G, and $N(T) \leqq M - 1$.

Proof (a). For any distinct $a, b \in v(G)$, T includes a path between a and b. Hence for any $l \in T$, $T + l$ includes a circuit, that is, T is maximal. (b). If T were not a tree, there would exist a tree T' including T as a proper subset. Then $N(T') > M - 1$ in contradiction to Theorem 2-5 (d). (c). We employ induction. Since $N(T) < M$, from (1-4) it is impossible that $D(a, T) \geqq 2$ for all vertices. Since $D(a, T) \neq 0$ for all vertices, there exists at least one vertex a such that $D(a, T) = 1$. Let $\{l\} = T \cap S[a]$; in G/l, $T - l$ satisfies all the assumptions of (c), whence by induction assumption it includes no circuit in G/l and therefore also in G because of Theorem 2-2 (c). Thus T includes no circuit in G. Hence Theorem 2-6 (*a*) implies that T is a tree. q.e.d.

THEOREM **2-7** Let G be a connected graph and $T \subset |G|$.

(a) If $I \cap T = \emptyset$ then $T \in \boldsymbol{T}_G$ implies $T \in \boldsymbol{T}_{G-I}$ and *vice versa.*

(b) If $I \subset T$ and if $T \in \boldsymbol{T}_G$ then $T - I \in \boldsymbol{T}_{G/I}$.

Proof The propositions can be verified by using Theorems 2-5 and 2-6. q.e.d.

THEOREM **2-8** For any two trees T and T' in a connected graph G, we can always find a sequence of trees $T = T_0, T_1, \ldots, T_{k-1}, T_k = T'$ such that

$$N(T_i \cap T_{i+1}) = M - 2 \quad (i = 0, 1, \ldots, k - 1) \tag{2-1}$$

Proof We employ induction. Suppose that we have found trees $T_0, T_1, \ldots, T_j$ satisfying (2-1) with j instead of k. If $T_j \neq T'$ then there exists $l \in T' - T_j$. Let P be a path included in T_j between the two end points of l [see Theorem 2-5 (c)]. Since $P + l$ is a circuit, we have $P \not\subset T'$, whence there exists $l' \in P - T'$. Then, since $T_j \in \boldsymbol{T}, T_{j+1} \equiv T_j + l - l'$ is also a tree because of Theorems 2-5 and 2-6, and

$$N(T_{j+1} \cap T') = N(T_j \cap T') + 1 \tag{2-2}$$

because $l \in T'$ but $l' \notin T'$. Thus we can proceed from $T_0 = T$ to T' in such a way that (2-1) is satisfied. q.e.d.

THEOREM **2-9** In a connected graph G, given a 2-tree T^2, there is a unique cut-set S such that $S \cap T^2 = \emptyset$.

Proof Since T^2 is a tree minus one line, the subgraph $G - (T^2)^*$ consists of two connected components because of Theorem 2-5 (a) (b) (c). Hence $(T^2)^*$ has to include a unique cut-set S. q.e.d.

We denote by $\boldsymbol{T}^2(S)$ the set of all 2-trees T^2 such that $S \cap T^2 = \emptyset$. Furthermore, we put

$$\boldsymbol{T}^2(h|h') \equiv \bigcap_{S \in \boldsymbol{S}(h|h')} \boldsymbol{T}^2(S) \tag{2-3}$$

Likewise $\boldsymbol{T}^2(a \mid b)$, etc. are defined; then, as seen from Theorem 2-6, $T^2 \in \boldsymbol{T}^2(a \mid b)$ is a tree in $G^{(ab)}$. In general, an n-tree T^n is a tree in a graph which is obtained from G by identifying n appropirate vertices.

A pseudotree T^0, which is a tree plus one line by definition includes a unique circuit because of Theorem 2-5 (c).

2-2 Fundamental Sets of Circuits and Cut-Sets

In order to discuss properties of a graph G algebraically, it is important to introduce some sign factors, called incidence numbers, circuit numbers, and cut-sets numbers.

DEFINITION **2-10** For any pair of a vertex a and a line l, we define an *incidence number* $[a:l]$ as follows:

$$\begin{aligned}[a:l] &= \pm 1 \quad \text{if } l \in S[a] - L[a] \\ &= 0 \quad \text{otherwise}\end{aligned} \tag{2-4}$$

Here the double sign is determined as follows: $[a:l] = +1$ if a is the outgoing end point of l and $[a:l] = -1$ if a is the ingoing one.

From the above definition, for any connected component G' of G we have

$$\sum_{a \in v(G')} [a, l] = 0 \quad \text{for any } l \in |G'| \tag{2-5}$$

because l has only one outgoing end point and only one ingoing one.

Circuits and cut-sets can be given orientations. Intuitively, an orientation of a circuit C is a way of going around C and that of a cut-set S is an order of two relevant connected subgraphs of $G - S$.

DEFINITION **2-11** For any pair of a circuit C and a line l, we define a *circuit number* $[C:l]$ as follows:

$$\begin{aligned}[C:l] &= \pm 1 \quad \text{if } l \in C \\ &= 0 \quad \text{otherwise}\end{aligned} \tag{2-6}$$

Here the double sign is relatively determined as follows: If $l \in C$ and $k \in C$ are both incident with a, then

$$[a:l]\,[C:l] = -\,[a:k]\,[C:k] \tag{2-7}$$

The overall sign of $[C:l]$ can be chosen arbitrarily.

We can rewrite (2-7) as

$$\sum_{l \in |G|} [a:l]\,[C:l] = 0 \tag{2-8}$$

because $[a:l]\,[C:l] \neq 0$ only for two lines since $D(a, C) = 2$.

DEFINITION **2-12** For any pair of a cut-set S and a line l, we define a *cut-set number* $[S : l]$ as follows:

$$[S:l] = \pm 1 \quad \text{if } l \in S$$
$$= 0 \quad \text{otherwise} \tag{2-9}$$

Here the double sign is relatively determined as follows: $[S : l]$ equals $[a_S, l]$, where a_S is one of the two non-isolated points in $G/(|G| - S)$. The overall sign of $[S : l]$ can be chosen arbitrarily.

THEOREM **2-13** Given an arbitrary cut-set S, let H be one of the two connected components of $G - S$ which are linked in G. Then

$$[S:l] = \pm \sum_{a \in v(H)} [a:l] \tag{2-10}$$

where the double sign is independent of l.

Proof If $l \notin S$ then *either* the two end points of l belong to H *or* neither of them belongs to H. In both cases, the right-hand side of (2-10) vanishes as required. If $l \in S$ then it equals $[a_S : l]$ defined above, because in general for the two end points a and b of a line k an incidence number of the vertex $a = b$ in G/k is a sum of the corresponding incidence number of a and that of b in G. q.e.d.

THEOREM **2-14** For any $C \in \boldsymbol{C}$ and any $S \in \boldsymbol{S}$, we have

$$N(C \cap S) = \text{even}, \tag{2-11}$$

$$\sum_{l \in |G|} [C:l]\,[S:l] = 0 \tag{2-12}$$

Proof From (2-10) and (2-8)

$$\sum_{l} [C:l]\,[S:l] = \pm \sum_{a \in v(H)} \sum_{l} [C:l]\,[a:l] = 0 \tag{2-13}$$

Since $|[C : l]\,[S : l]| = 1$ if and only if $l \in C \cap S$, (2-11) follows from (2-12). q.e.d.

DEFINITION **2-15** Circuits $C_1, C_2, \ldots, C_k$ are called *independent* if the quantities $\sum_{l \in |G|} [C_j : l]\,\alpha_l$ $(j = 1, 2, \ldots, k)$ are linearly independent for any arbitrary α_l $(l \in |G|)$, that is, if the relations

$$\sum_{j=1}^{k} e_j[C_j : l] = 0 \tag{2-14}$$

do not hold for all $l \in |G|$ simultaneously for any set of real numbers $e_1, e_2, \ldots, e_k$ such that $\sum_{j=1}^{k} |e_j| \neq 0$.

THEOREM **2-16** All maximal sets of independent circuits have the same number of elements.

Proof Let $\{C_1, C_2, \ldots, C_\mu\}$ be a maximal set of independent circuits which consists of the minimum number of elements. By definition, for any circuits C'_k we can always find real numbers $e_{k1}, e_{k2}, \ldots, e_{k\mu}$ such that

$$[C'_k : l] = \sum_{j=1}^{\mu} e_{kj}[C_j : l] \tag{2-15}$$

for all $l \in |G|$. Suppose that $C'_1, C'_2, \ldots, C'_{\mu+1}$ were independent. Then, since $C'_1, C'_2, \ldots, C'_\mu$ are independent, $\det(e_{kj} \mid k, j = 1, \ldots, \mu)$ cannot vanish. Furthermore, not all of $e_{\mu+1,j}$ $(j = 1, \ldots, \mu)$ are zero. Hence we can always find e'_k $(k = 1, 2, \ldots, \mu + 1)$ $(\sum_{k=1}^{\mu+1} |e'_k| \neq 0)$ such that

$$\sum_{k=1}^{\mu+1} e'_k e_{kj} = 0 \quad (j = 1, \ldots, \mu) \tag{2-16}$$

Hence (2-15) implies

$$\sum_{k=1}^{\mu+1} e'_k[C'_k : l] = 0 \tag{2-17}$$

that is, $C'_1, C'_2, \ldots, C'_{\mu+1}$ are not independent. q.e.d.

DEFINITION **2-17** The number of elements of a maximal set of independent circuits is called the *connectivity* (or *nullity*, or *first Betti number*). We denote the connectivity of G by $\mu(G)$ or simply by μ. The connectivity $\mu(I)$ of $I \subset |G|$ is defined by $\mu(H)$, where H is a subgraph of G such that $|H| = I$.

Note In an unoriented graph, independent circuits are defined as follows. Since for any circuits C_1 and C_2, the symmetric difference

$$C_1 \oplus C_2 \equiv C_1 \cup C_2 - C_1 \cap C_2 \tag{2-18}$$

is a disjoint union of circuits, $C_1 \oplus C_2 \oplus \ldots \oplus C_k$ is also a disjoint union of circuits. Then $C_1, C_2, \ldots, C_k$ are called independent if any symmetric difference made out of them is not empty. We note that if $C_1, C_2, \ldots, C_k$ are independent in this sense then they are also independent in the sense of

Definition 2-15, but the converse is not true. For example, in Fig. 2-1, circuits $C_1 = \{1, 2, 3, 4\}$, $C_2 = \{1, 5, 3, 6\}$, $C_3 = \{2, 6, 4, 5\}$ are independent in the sense of Definition 2-15 but not of the definition presented here. The connectivity is, however, the same in both definitions.

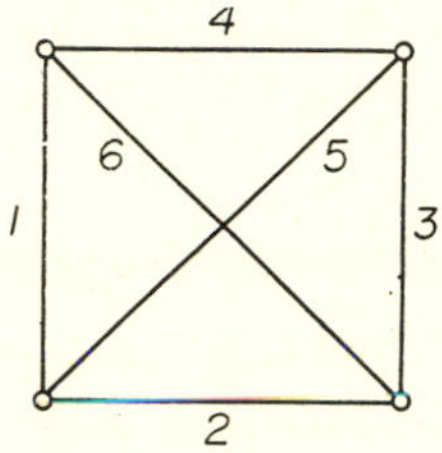

Fig. 2-1 A graph having three independent circuits.

DEFINITION **2-18** If a set of μ circuits $C_1, C_2, \dots, C_\mu$ satisfies the condition

$$C_k \not\subset \bigcup_{j \neq k} C_j \quad (k = 1, 2, \dots, \mu) \tag{2-19}$$

then it is called a *fundamental set of circuits*, or simply an *f-set of circuits*.†

By definition, it is obvious that if $\{C_1, \dots, C_\mu\}$ is an f-set of circuits then $C_1, \dots, C_\mu$ are independent, but the converse is not true. For example, in Fig. 2-1, $\{C_1, C_2, C\}$ is not an f-set of circuits for any $C \in \mathbf{C}$. We also note that every circuit consisting of a loop line alone belongs to any f-set of circuits.

THEOREM **2-19** If $\{C_1, \dots, C_\mu\}$ is an f-set of circuits, then for any circuit C we have

$$[C : l] = \sum_{j=1}^{\mu} [C_j : C]\,[C_j : l] \tag{2-20}$$

where $[C_j : C] = \pm 1$ or 0 (independently of l).

Proof Since $C_1, \dots, C_\mu$ are independent, we can write

$$[C : l] = \sum_{j=1}^{\mu} e_j [C_j : l] \tag{2-21}$$

Choosing $l = l_j \in C_j - \bigcup_{k \neq j} C_k$, we find

$$[C : l_j] = e_j [C_j : l_j] \tag{2-22}$$

with $[C_j : l_j] = \pm 1$. Thus we can rewrite (2-21) as (2-20).

† The existence of f-sets of circuits follows from Theorem 2-22 below.

THEOREM **2-20** If $\{C_1, \ldots, C_\mu\}$ is an f-set of circuits in a connected graph, then $T^* \equiv \{l_1, l_2, \ldots, l_\mu\}$ is a chord set, where $l_j \in C_j - \bigcup_{k \neq j} C_k$ $(j = 1, 2, \ldots, \mu)$.

Proof Let $T \equiv |G| - T^*$. For any $C \in \mathbf{C}$, (2-21) with (2-22) implies that there exists l_j such that $l_j \in C$, whence T includes no circuit because $l_j \notin T$. Furthermore $T + l_j \supset C_j$ because $T^* \cap C_j = \{l_j\}$, whence T is maximal. Thus T is a tree. q.e.d.

THEOREM **2-21** If G is a graph consisting of p connected components, the connectivity μ of G is given by

$$\mu = N - M + p \tag{2-23}$$

Proof If G is connected ($p = 1$), Theorems 2-20 and 2-5(d) imply $\mu = N - M + 1$. In the general case, this relation holds for each connected component, whence we have (2-23). q.e.d.

THEOREM **2-22** Given a chord set T^* in a connected graph, for any $l_j \in T^*$ there exists a unique circuit C_j such that $T^* \cap C_j = \{l_j\}$, and therefore T^* uniquely determines an f-set of circuits $\{C_1, \ldots, C_\mu\}$.

Proof Because of Theorem 2-5 (c), for any $l_j \in T^*$ there exists uniquely a path $P_j \subset T$ such that $P_j + l_j \equiv C_j$ is a circuit. q.e.d.

We can discuss the cut-sets in quite an analogous way to the above discussion on the circuits. Accordingly, we sometimes omit proofs.

DEFINITION **2-23** Cut-sets $S_1, S_2, \ldots, S_k$ are called *independent* if the quantities $\sum_{l \in |G|} [S_j : l]\, \beta_l$ $(j = 1, 2, \ldots, k)$ are linearly independent for any arbitrary β_l $(l \in |G|)$.

DEFINITION **2-24** Since the number of elements of a maximal set of independent cut-sets is independent of the choice of cut-sets as before, it is called the *rank*, and usually denoted by $r(G)$ or simply by r.

DEFINITION **2-25** If a set $\{S_1, \ldots, S_r\}$ of r (independent) cut-sets satisfies the condition

$$S_k \not\subset \bigcup_{j \neq k} S_j \, (k = 1, 2, \ldots, r) \tag{2-24}$$

then it is called a *fundamental set of cut-sets*, or simply an *f-set of cut-sets*.

It is evident that every cut-set consisting of a cut-line alone belongs to any f-set of cut-sets.

THEOREM **2-26** If $\{S_1, \ldots, S_r\}$ is an f-set of cut-sets, then for any cut-set S we have

$$[S:l] = \sum_{j=1}^{r} [S_j:S]\,[S_j:l] \tag{2-25}$$

where $[S_j:S] = \pm 1$ or 0 (independently of l).

THEOREM **2-27** If $\{S_1, \ldots, S_r\}$ is an f-set of cut-sets in a connected graph, then $T \equiv \{l_1, \ldots, l_r\}$ is a tree, where $l_j \in S_j - \bigcup_{k \neq j} S_k$ $(j = 1, 2, \ldots, r)$.

Proof If T included a circuit C, then one would have $C \cap S_j \neq \emptyset$ for some S_j. Hence (2-11) implies

$$N(T \cap S_j) \geqq N(C \cap S_j) \geqq 2 \tag{2-26}$$

in contradiction with $N(T \cap S_j) = 1$ (definition of T). Hence we have only to show that $T + l$ for any $l \notin T$ includes a circuit. Among $S_1, \ldots, S_r$, let $S_1, \ldots, S_m$ $(m \leqq r)$, say, be cut-sets which contain l. Then for $I \equiv \{l, l_1, \ldots, l_m\}$ we see $N(I \cap S_j) = 2$ for $j = 1, \ldots, m$ and $N(I \cap S_j) = 0$ for $j = m + 1, \ldots, r$. Hence for any $S \in \mathbf{S}$, we have $N(I \cap S) =$ even, whence in particular $N(I \cap S[a]) =$ even for any vertex a because of Theorem 2-3(b). Thus $I(\subset T + l)$ includes a circuit. q.e.d.

THEOREM **2-28** If G is a graph consisting of p connected components, the rank r of G is given by

$$r = M - p \tag{2-27}$$

namely

$$\mu + r = N \tag{2-28}$$

THEOREM **2-29** Given a tree T in a connected graph G, for any $l_j \in T$ there exists a unique cut-set S_j such that $T \cap S_j = \{l_j\}$, and therefore T uniquely determines an f-set of cut-sets $\{S_1, \ldots, S_r\}$.

Proof Since the reduced graph $G/(T - l_j)$ has only two vertices, it contains a unique cut-set S_j (of course $l_j \in S_j$). Then S_j is a cut-set also in G because of Theorem 2-2(d). q.e.d.

THEOREM **2-30** In a nonseparable graph G with $M \geqq 2$, M $-$ 1 stars are independent cut-sets.

Proof Any star in G is a cut-set because of Theorem 2-3(a). Any cut-set is expressible in terms of $M - 1$ stars in the sense of Theorem 2-13. If $M - 1$ stars were not independent, therefore, the maximal number of independent cut-sets would be less than $r = M - 1$. q.e.d.

2-3 Duality

Within each of the following pairs, one is called the dual concept of the other:

circuit	$\leftrightarrow$	cut-set
chord set	$\leftrightarrow$	tree
contraction	$\leftrightarrow$	deletion
connectivity	$\leftrightarrow$	rank
loop line	$\leftrightarrow$	cut-line

If a proposition Γ, which does not contain the concepts of vertex, star, path, etc., is true, then the dual proposition $\tilde{\Gamma}$, which is obtained from Γ by replacing the concepts in Γ by their duals, is also true. For example, we have the following pairs: Theorem 2-1(c) $\leftrightarrow$ Theorem 2-2(d), Theorem 2-2(c) $\leftrightarrow$ Theorem 2-1(d), Theorem 2-19 $\leftrightarrow$ Theorem 2-26, Theorem 2-20 $\leftrightarrow$ Theorem 2-27, Theorem 2-21 $\leftrightarrow$ Theorem 2-28, Theorem 2-22 $\leftrightarrow$ Theorem 2-29.

In Section 4, dual graphs will be introduced only for planar graphs, but the duality defined above is meaningful for any graphs. When graph theory is applied to the Feynman integral in Chapter 2, the duality will play an important role, that is, it is closely related to the reciprocity between space-time position and energy-momentum.

§ 3 MATRICES AND TOPOLOGICAL FORMULAS

3-1 Matrices Associated with a Graph

In order to deal with algebraic quantities associated with a graph G systematically, we introduce some matrices, called the incidence (or vertex) matrix, the circuit matrix, and the cut-set matrix. Matrices are always denoted by script letters. We denote the unit matrix by $\mathscr{E}$ and the zero matrix by $\mathscr{O}$; if necessary, the unit matrix of order k and the $m \times n$ zero matrix are denoted by $\mathscr{E}^{(k)}$ and by $\mathscr{O}^{(m \times n)}$, respectively.

DEFINITION **3-1** The *incidence matrix* of G is an $M \times N$ matrix defined by

$$\mathscr{A} \equiv ([a : l] \mid a \in v(G),\ l \in |G|) \tag{3-1}$$

The *circuit matrix* of G is defined by

$$\mathscr{C} \equiv ([C : l] \mid C \in \boldsymbol{C},\ l \in |G|) \tag{3-2}$$

The *cut-set matrix* of G is defined by

$$\mathscr{S} \equiv ([S : l] \mid S \in \boldsymbol{S},\ l \in |G|) \tag{3-3}$$

We note that $\mathscr{A}$ is essentially† a submatrix of $\mathscr{S}$ if G is nonseparable and $M \geqq 2$ because of Theorem 2-3.

THEOREM **3-2** The ranks of the matrices, $\mathscr{A}$, $\mathscr{C}$, and $\mathscr{S}$ are as follows:

$$\text{rank } \mathscr{A} = r \tag{3-4}$$

$$\text{rank } \mathscr{C} = \mu \tag{3-5}$$

$$\text{rank } \mathscr{S} = r \tag{3-6}$$

[Here, as is well known, the rank of a matrix is the maximum order of its square submatrices whose determinants are nonzero.]

Proof The relations (3-5) and (3-6) are direct consequences of the definitions of μ and r, respectively; (3-4) follows from (2-5) and (2-10) together with (3-6). q.e.d.

From Definition 3-1, we see that $\mathscr{A}$, $\mathscr{C}$, and $\mathscr{S}$ are direct sums of their submatrices corresponding to connected components of G. Without loss of

† For $\mathscr{A}$ to be a true submatrix of $\mathscr{S}$, one has to assume that in Definition 2-12 if $S = S[a]$ then a_S is chosen to be a, and moreover for $M = 2$, the row of $\mathscr{S}$ should be doubled.

generality, we may hereafter confine ourselves to considering a connected graph G. Then $r = M - 1$. Therefore, it is convenient to consider a submatrix $\mathscr{A}^a$ of $\mathscr{A}$, which is obtained from $\mathscr{A}$ by deleting a row corresponding to a particular vertex a. Furthermore, for $I \subset |G|$, we denote by $\mathscr{A}_I^a$ the submatrix of $\mathscr{A}^a$ whose columns are only the ones corresponding to the lines belonging to I.

THEOREM **3-3** Let T be a subset of $|G|$ such that $N(T) = M - 1$. Then $\det \mathscr{A}_T^a$ is nonzero if and only if T is a tree, and in that case

$$\det \mathscr{A}_T^a = \pm 1 \tag{3-7}$$

Proof If T is not a tree, it includes a circuit C because of Theorem 2-6(b). Then (2-8) implies that the columns corresponding to the lines $l \in C$ are not linearly independent. Thus $\det \mathscr{A}_T^a = 0$. Now, let T be a tree. We employ induction. Let $l \in T \cap S[a]$; then the column corresponding to l has only one nonzero element in $\mathscr{A}_T^a$. Therefore, expanding $\det \mathscr{A}_T^a$, apart from a sign factor it reduces to $\det \mathscr{A}_{T-l}^{a,b}$, where b is the other end point of l and $\mathscr{A}_I^{a,b}$ denotes a submatrix of $\mathscr{A}$ which is obtained by deleting two rows a and b and the columns of lines not contained in I. Since $T - l$ is a tree in G/l because of Theorem 2-7(b), we have $\det \mathscr{A}_{T-l}^{a,b} = \pm 1$ by induction assumption because $a = b$ in G/l. [(3-7) is trivially true in the case $N(T) = 1$.] q.e.d.

We denote the transposition of a matrix by affixing a superscript t. Then (2-8) and (2-12) can be rewritten in the following matrix forms.

THEOREM **3-4** One has orthogonality relations

$$\mathscr{C} \mathscr{A}^t = \mathscr{O} \tag{3-8}$$

$$\mathscr{C} \mathscr{S}^t = \mathscr{O} \tag{3-9}$$

We denote by C_{T*} the $\mu \times N$ submatrix of $\mathscr{C}$ whose rows correpond to the f-set of circuits determined by a chord set T^* in the sense of Theorem 2-22. Likewise, we denote by $\mathscr{S}_T$ the $r \times N$ submatrix of $\mathscr{S}$ whose rows correspond to the f-set of cut-sets determined by a tree T in the sense of Theorem 2-29.

THEOREM **3-5** We can array the columns of $\mathscr{C}_{T*}$ and $\mathscr{S}_T$ in such a way that

$$\mathscr{C}_{T*} = [\mathscr{E}^{(\mu)}, \mathscr{B}_T] \tag{3-10}$$

$$\mathscr{S}_T = [-(\mathscr{B}_T)^t, \mathscr{E}^{(r)}] \tag{3-11}$$

where $\mathscr{B}_T$ is a $\mu \times r$ matrix ($r = M - 1 = N - \mu$).

Proof Let $T^* = \{l_1, l_2, \ldots, l_\mu\}$ and $T = \{l_{\mu+1}, \ldots, l_N\}$. If $\{C_1, \ldots, C_\mu\}$ and $\{S_{\mu+1}, \ldots, S_N\}$ are the f-set of circuits corresponding to T^* and that of cut-sets corresponding to T, respectively, then we have

$$[C_i : l_j] = \delta_{ij} \quad (i, j = 1, \ldots, \mu)$$
$$[S_i : l_j] = \delta_{ij} \quad (i, j = \mu + 1, \ldots, N) \tag{3-12}$$

by choosing the overall signs of $[C_i : l]$ and $[S_i : l]$ in such a way that $[C_i : l_i] > 0$ and $[S_i : l_i] > 0$. Thus we can write $\mathscr{C}_{T*}$ as (3-10) and $\mathscr{S}_T = [\mathscr{B}'_T, \mathscr{E}^{(r)}]$. Then Theorem 3-4 implies

$$\mathscr{C}_{T*}(\mathscr{S}_T)^t = (\mathscr{B}'_T)^t + \mathscr{B}_T = \mathscr{O} \tag{3-13}$$

Hence we have (3-11). q. e. d.

The following theorem is evident from the definitions of the connectivity and the rank.

THEOREM **3-6** Let $\mathscr{C}_1$ and $\mathscr{C}_2$ be two submatrices of $\mathscr{C}$ each of which has μ rows corresponding to μ independent circuits. Then there exists a nonsingular matrix $\mathscr{J}$ of order μ such that $\mathscr{C}_2 = \mathscr{J}\mathscr{C}_1$. The corresponding proposition is true also for the cut-set matrix.

THEOREM **3-7** Let T and T' be two trees; then

$$\mathscr{C}_{T'*} = \mathscr{J}\,\mathscr{C}_{T*} \qquad \text{with } \det \mathscr{J} = \pm 1 \tag{3-14}$$
$$\mathscr{S}_{T'} = \tilde{\mathscr{J}}\,\mathscr{S}_T \qquad \text{with } \det \tilde{\mathscr{J}} = \pm 1 \tag{3-15}$$

Proof The existence of nonsingular matrices $\mathscr{J}$ and $\tilde{\mathscr{J}}$ follows from Theorem 3-6. Because of Theorem 2-8 we may assume $N(T \cap T') = M - 2$ without loss of generality. Hence we can write the chord sets T^* and T'^* as

$$\begin{aligned} T^* &= \{l_1, \ldots, l_{\mu-1}, l_\mu\} \\ T'^* &= \{l_1, \ldots, l_{\mu-1}, l_{\mu+1}\} \end{aligned} \tag{3-16}$$

We array $\mathscr{C}_{T*}$ in the form of (3-10). Then $\mathscr{J}$ is given by the first μ columns of $\mathscr{C}_{T'*}$. By the definition of $\mathscr{C}_{T'*}$, $\mathscr{J}$ has the following form:

$$\mathscr{J} = \left(\begin{array}{cccc|c} 1 & 0 & \cdots & 0 & * \\ 0 & 1 & \cdots & 0 & * \\ \vdots & \vdots & & \vdots & \vdots \\ 0 & 0 & \cdots & 1 & * \\ \hline 0 & 0 & \cdots & 0 & \pm 1 \end{array}\right) \tag{3-17}$$

where * stands for 0 or ± 1. Hence we have (3-14).

We can likewise prove (3-15). q.e.d.

THEOREM **3-8** In the array of (3-11), one has

$$\mathscr{S}_T = (\mathscr{A}_T^a)^{-1}\,\mathscr{A}^a \tag{3-18}$$

namely

$$(\mathscr{B}_T)^t = -\,(\mathscr{A}_T^a)^{-1}\mathscr{A}_{T*}^a \tag{3-19}$$

Proof Any row of $\mathscr{A}$ can be expressed as a linear combination of rows of $\mathscr{S}_T$ because of Theorem 2-3 together with Definition 2-12 and Theorem 2-26. Hence there exists a matrix $\mathscr{K}$ such that

$$\mathscr{A}^a = \mathscr{K}\,\mathscr{S}_T \tag{3-20}$$

On substituting (3-11) for $\mathscr{S}_T$ in (3-20), we find $\mathscr{K} = \mathscr{A}_T^a$. Since $\mathscr{A}_T^a$ is nonsingular because of (3-7), (3-18) follows from (3-20). q.e.d.

3-2 Topological Formulas

In this subsection, we calculate determinants of some matrices associated with a connected graph G. Since they are expressed as algebraic formulas on which some topological structures of G reflect directly, those formulas are called *topological formulas.*

Let $\{\alpha_l \mid l \in |G|\}$ and $\{\beta_l \mid l \in |G|\}$ be two sets of N parameters. When applied to the theory of the Feynman integral, α_l and β_l are a *Feynman parameter* and an *inverse Feynman parameter*, respectively. Topological formulas will become important in the Feynman-parametric integral and the inverse-Feynman-parametric integral, as seen in Chapter 2.

Let $\mathscr{I}$ and $\tilde{\mathscr{I}}$ be two diagonal matrices defined by

$$\mathscr{F} \equiv (\alpha_l\delta_{lk} \mid l, k \in |G|)$$

$$\tilde{\mathscr{F}} \equiv (\beta_l\delta_{lk} | l, k \in |G|) \tag{3-21}$$

where δ_{lk} is the Kronecker delta ($\delta_{ll} = 1$ and $\delta_{lk} = 0$ for $l \neq k$). Furthermore, we define the following matrices:

$$\mathscr{W}^a \equiv \mathscr{A}^a\,\tilde{\mathscr{F}}(\mathscr{A}^a)^t = \Big(\sum_{l\in|G|} [b:l]\,[b':l]\,\beta \,|b, b' \in v(G) - a\Big) \tag{3-22}$$

$$\mathscr{U}_{T*} \equiv \mathscr{C}_{T*}\mathscr{F}(\mathscr{C}_{T*})^t = \Big(\sum_{l\in|G|} [C_i:l]\,[C_j:l]\,\alpha \,|i, j = 1, \ldots, \mu\Big) \tag{3-23}$$

$$\tilde{\mathscr{U}}_T \equiv \mathscr{S}_T\tilde{\mathscr{F}}(\mathscr{S}_T)^t = \Big(\sum_{l\in|G|} [S_i:l]\,[S_j:l]\,\beta \,|i, j = \mu + 1, \ldots, N\Big) \tag{3-24}$$

where $\{C_1, \ldots, C_\mu\}$ and $\{S_{\mu+1}, \ldots, S_N\}$ are the *f*-set of circuits corresponding to a chord set T^* and that of cut-sets corresponding to a tree T, respectively.

DEFINITION **3-9** A homogeneous polynomial of degree μ,

$$U(\alpha) \equiv \sum_{T^* \in T^*} \prod_{l \in T} \alpha_l \tag{3-25}$$

is called the *chord-set product sum*, and a homogeneous polynomial of degree $M - 1$,

$$\tilde{U}(\beta) \equiv \sum_{T \in T} \prod_{l \in T} \beta_l \tag{3-26}$$

is called the *tree product sum*.

Example For Fig. 3-1, we have

$$U(\alpha) = (\alpha_1 + \alpha_2 + \alpha_3)\,\alpha_4 + \alpha_4(\alpha_5 + \alpha_6 + \alpha_7) + (\alpha_5 + \alpha_6 + \alpha_7)(\alpha_1 + \alpha_2 + \alpha_3)$$

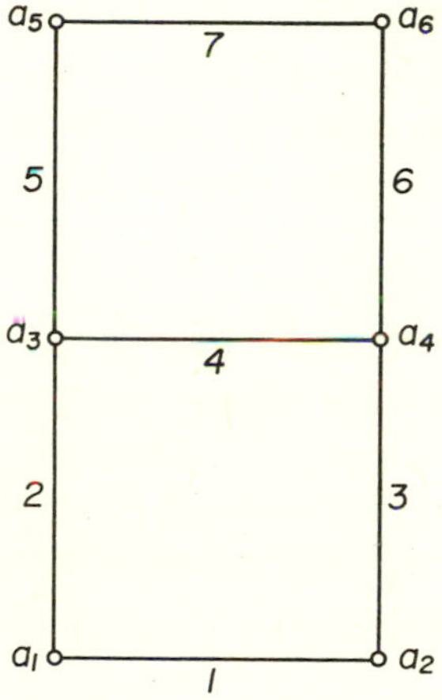

Fig. 3-1 A graph with $N = 7$, $M = 6$, and $\mu = 2$.

From the definition, it is evident that

$$\tilde{U}(1/\alpha) = U(\alpha)/\prod_{l \in |G|} \alpha_l \tag{3-27}$$

$$U(1/\beta) = \tilde{U}(\beta)/\prod_{l \in |G|} \beta_l \tag{3-28}$$

THEOREM **3-10**

$$\det \mathscr{W}^a = \det \tilde{\mathscr{U}}_T = \tilde{U} \tag{3-29}$$

$$\det \mathscr{U}_{T^*} = U \tag{3-30}$$

[Note that those determinants are independent of a and T.]

Proof In the literatures of graph theory [1, 4] it is customary to use the Binet–Cauchy theorem [7] for a determinant of a product of two rectangular matrices. According to it, one has

$$\det \mathscr{W}^a = \sum_{N(T)=M-1} \det(\mathscr{A}^a_T \tilde{\mathscr{F}}_T) \det \mathscr{A}^a_T$$
$$= \sum_{N(T)=M-1} (\det \mathscr{A}^a_T)^2 \prod_{l\in T} \beta_l \tag{3-31}$$

where $\tilde{\mathscr{I}}_T$ is the restriction of $\tilde{\mathscr{I}}$ to T. Then because of Theorem 3-3 we immediately have $\det \mathscr{W}^a = \tilde{U}$. Here, we present another proof, however.

We first prove (3-30) [8]. From (3-14), we observe that $\det \mathscr{U}_{T*}$ is independent of the choice of T^*. From (3-10) we see that for a particular T^*, $\mathscr{U}_{T*}$ contains $\alpha_l (l \in T^*)$ only in its principal diagonal line and only once. Hence $\det \mathscr{U}_{T*}$ contains $\prod_{l\in T*} \alpha_l$ for any chord set T^*. Let

$$\det \mathscr{U}_{T*} = U + R \tag{3-32}$$

then from the above reasoning R is a sum of products $\alpha_{l_1}\alpha_{l_2} \ldots \alpha_{l_\mu}$ such that $\{l_1, l_2, \ldots, l_\mu\}$ is not a chord set. We assume that R contains a term of the form $k\, \alpha_{l_1}\alpha_{l_2} \ldots \alpha_{l_\mu}$ $(k \neq 0)$ for some $l_1, l_2, \ldots, l_\mu$, which may not be all distinct, and show absurdity. Let $I \equiv \bigcup_{j=1}^{\mu} \{l_j\}$; then $N(I^*) \geqq M - 1$ and $I^* \equiv |G| - I$ is not a tree. Hence because of Theorem 2-6(b) I^* includes a circuit C. By definition, any chord set T^* intersects any circuit, whence $T^* \cap C \neq \emptyset$. Put $\alpha_l = 0$ for all $l \in C$. Then $U = 0$, and $\det \mathscr{U}_{T*} = 0$ because all elements of the row C of $\mathscr{U}_{T*}$ vanish if T^* is chosen† in such a way that C belongs to the f-set of circuits corresponding to T^*. However, $R \neq 0$ when $\alpha_l = 0$ for all $l \in C$ because of $I \cap C = \emptyset$; this is a contradiction. Thus $R \equiv 0$.

We can likewise prove the second equality of (3-29). An explicit proof of it from (3-30) will also be presented below. Finally, the first equality of (3-29) follows from Theorem 3-8 together with (3-7). q.e.d.

THEOREM **3-11**

$$\det \begin{bmatrix} \mathscr{C}_{T*}\mathscr{F} \\ \mathscr{S}_T \end{bmatrix} = U \tag{3-33}$$

$$\det \begin{bmatrix} \mathscr{C}_{T*}\mathscr{F} \\ \mathscr{A}^a \end{bmatrix} = \pm\, U \tag{3-34}$$

† For example, if $l_1 \in C$ and if $\{l_2, \ldots, l_\mu\}$ is a chord set in G/C, then $T^* = \{l_1, \ldots, l_\mu\}$.

Proof Let $\mathscr{E}_T \equiv [\mathcal{O}^{(r\times\mu)}, \mathscr{E}^{(r)}]$. Then (3-10) leads to

$$\det [(\mathscr{C}_{T*})^t, (\mathscr{E}_T)^t] = \det \begin{bmatrix} \mathscr{E}^{(\mu)}, \mathcal{O}^{(\mu\times r)} \\ (\mathscr{B}_T)^t, \mathscr{E}^{(r)} \end{bmatrix} = 1 \tag{3-35}$$

Since

$$\begin{bmatrix} \mathscr{C}_{T*}\mathscr{F} \\ \mathscr{S}_T \end{bmatrix} [(\mathscr{C}_{T*})^t, (\mathscr{E}_T)^t] = \begin{bmatrix} \mathscr{U}_{T*}, & \mathscr{C}_{T*}\mathscr{F}(\mathscr{E}_T)^t \\ \mathcal{O}^{(r\times\mu)}, & \mathscr{E}^{(r)} \end{bmatrix} \tag{3-36}$$

because of (3-23), (3-9), and (3-11), we find

$$\det \begin{bmatrix} \mathscr{C}_{T*}\mathscr{F} \\ \mathscr{S}_T \end{bmatrix} = \det \mathscr{U}_{T*} = U \tag{3-37}$$

with the aid of (3-30). The second formula (3-34) can likewise be proven by means of (3-7). q.e.d.

This theorem will become important for the analysis of the Landau equations presented in Section 12. Furthermore, we can derive (3-29) from (3-30) by means of Theorem 3-11. Put $\beta_l = 1/\alpha_l$ ($l \in |G|$) so that $\tilde{\mathscr{F}} = \mathscr{F}^{-1}$. Since we can analogously prove

$$\det \tilde{\mathscr{U}}_T = \det \begin{bmatrix} \mathscr{C}_{T*} \\ \mathscr{S}_T\tilde{\mathscr{F}} \end{bmatrix} \tag{3-38}$$

we have

$$\det \tilde{\mathscr{U}}_T\big|_{\beta=1/\alpha} = \det \begin{bmatrix} \mathscr{C}_{T*}\mathscr{F} \\ \mathscr{S}_T \end{bmatrix} \det \tilde{\mathscr{F}}\big|_{\beta=1/\alpha}$$

$$= U/\prod_{l\in|G|} \alpha_l \tag{3-39}$$

Definition **3-12** For $C \in \boldsymbol{C}_G$, we denote by $U_C(\alpha)$ the chord-set product sum in G/C. If $\mu = 1$ then $U_C(\alpha) \equiv 1$.

Example For Fig. 3-1, for $C = \{4, 5, 7, 6\}$

$$U_C = \alpha_1 + \alpha_2 + \alpha_3$$

We denote the (C, C') cofactor of $\mathscr{U}_{T*}$ by $[\mathscr{U}_{T*}]^{C,C'}$; $[\mathscr{W}^a]^{b,b'}$ and $[\tilde{\mathscr{U}}_T]^{S,S'}$ are similar. We write $\alpha(I) = \{\alpha_l \mid l \in I\}$ and $\alpha(I) = 0$ means $\alpha_l = 0$ for all $l \in I$; $\beta(I)$ is similar.

Theorem **3-13**

$$[\mathscr{U}_{T*}]^{C,C}\big|_{\alpha(C)=0} = U_C \tag{3-40}$$

Proof Let $\{C, C_1, \ldots, C_{\mu-1}\}$ be the f-set of circuits in G which is determined by the chord set T^* in the sense of Theorem 2-22. Then it is straightforward to show that $\{C_1 - C, \ldots, C_{\mu-1} - C\}$ is the f-set of circuits corresponding to $T^* - C$ in G/C. The determinant of its $\mathscr{U}$-matrix is nothing but the left-hand side of (3-40). Hence it equals U_C by (3-30). q.e.d.

DEFINITION **3-14** Homogenous polynomials of degree $M - 2$,

$$\tilde{W}_S(\beta) \equiv \sum_{T^2 \in T^2(S)} \prod_{l \in T^2} \beta_l \tag{3-41}$$

$$\tilde{W}^{(h|h')}(\beta) \equiv \sum_{T^2 \in T^2(h|h')} \prod_{l \in T^2} \beta_l = \tilde{W}^{(h'|h)} \tag{3-42}$$

are called the *2-tree product sums* corresponding to a cut-set S and corresponding to a division $(h \mid h')$, respectively,† where h and h' are disjoint subsets of $v(G)$.

If $S \neq S'$ then $T^2(S)$ and $T^2(S')$ are disjoint because $G - S \cup S'$ consists of more than two connected components. Hence $\tilde{W}_S$ and $\tilde{W}_{S'}$ have no common terms for $S \neq S'$. Accordingly, we may rewrite (3-42) as

$$\tilde{W}^{(h|h')} = \sum_{S \in \boldsymbol{S}(h|h')} \tilde{W}_S \tag{3-43}$$

The following properties are direct consequences of Definition 3-14 with (3-43). For $u^* = v(G) - u$,

$$\begin{aligned} \tilde{W}^{(u|u^*)} &= \tilde{W}_S \quad \text{if } S \in \boldsymbol{S}(u|u^*) \\ &= 0 \qquad \text{if } \boldsymbol{S}(u|u^*) = \emptyset \end{aligned} \tag{3-44}$$

If $h \cap h' = \emptyset$ and $a \notin h \cup h'$, then

$$\tilde{W}^{(h|h')} = \tilde{W}^{(ha|h')} + \tilde{W}^{(h|h'a)} \tag{3-45}$$

where ha is an abbreviation of $h + a$. More generally, if $k, k' \subset g$ and $k \cap k' = \emptyset$,

$$\tilde{W}^{(k|k')} = \sum_h \tilde{W}^{(h|g-h)} \tag{3-46}$$

where the summation goes over all possible h such that $k \subset h$ and $k' \subset g - h$.

THEOREM **3-15**

$$[\tilde{\mathscr{U}}_T]^{S,S}|_{\beta(S)=0} = \tilde{W}_S \tag{3-47}$$

† See the end of Subsection 2-1.

Proof Let $\{S, S_1, \ldots, S_{M-2}\}$ be the f-set of cut-sets in G which is determined by the tree T in the sense of Theorem 2-29. By definition, $T \cap S = \{l\}$ and $l \notin S_j$ $(j = 1, 2, \ldots, M-2)$. Because of Theorems 2-2(d) and 2-7(b), $\{S_1, \ldots, S_{M-2}\}$ is the f-set of cut-sets corresponding to a tree $T - l$ in G/l. The determinant of its $\tilde{\mathscr{U}}$-matrix is nothing but $[\tilde{\mathscr{U}}_T]^{S,S}$, which equals the tree product sum in G/l because of (3-29). Since any tree in G/l which does not intersect $S - l$ is a 2-tree in G which does not intersect S, and *vice versa*, we obtain (3-47). q.e.d.

THEOREM **3-16**

$$[\mathscr{W}^a]^{b,b} = \tilde{W}^{(a|b)} \tag{3-48}$$

Proof By definition, $[\mathscr{W}^a]^{b,b}$ is nothing but the (ab, ab) minor, $[\mathscr{W}]^{ab,ab}$, of $\mathscr{W} \equiv \mathscr{A}\tilde{\mathscr{F}}\mathscr{A}^t$, that is, it coincides with the determinant of the $\mathscr{W}^a$ matrix in $G^{(ab)}$ because $a = b$ in $G^{(ab)}$. Hence (3-29) implies that $[\mathscr{W}^a]^{b,b}$ is the tree product sum in $G^{(ab)}$. Since a tree in $G^{(ab)}$ is equivalent to a 2-tree in G belonging to $T^2(a \mid b)$, we obtain (3-48) by Definition 3-14. q.e.d.

DEFINITION **3-17** In analogy with (3-27) and (3-28), we define

$$\tilde{U}_C(\beta) \equiv U_C(1/\beta) \prod_{l \in |G|} \beta_l \tag{3-49}$$

$$W_S(\alpha) \equiv \tilde{W}_S(1/\alpha) \prod_{l \in |G|} \alpha_l \tag{3-50}$$

$$W^{(h|h')}(\alpha) \equiv \tilde{W}^{(h|h')}(1/\alpha) \prod_{l \in |G|} \alpha_l \tag{3-51}$$

From the definitions, we immediately have

$$\tilde{U}_C = \sum_{T^0 \supset C} \prod_{l \in T^0} \beta_l \tag{3-52}$$

where the summation runs over all pseudotrees T^0 including C, and

$$W_S = U_H U_{H'} \prod_{l \in S} \alpha_l \tag{3-53}$$

where U_H and $U_{H'}$ denote the chord-set product sums of two connected components of G–S. Furthermore, $W^{(a|b)}$ is the chord-set product sum in $G^{(ab)}$, and the formulas without tildas corresponding to (3–43)–(3–46) hold.

Example For Fig. 3-1, for $S = \{5, 6\}$

$$W_S = (\alpha_1 + \alpha_2 + \alpha_3 + \alpha_4)\,\alpha_5\alpha_6$$

and if $a = a_1$ and $b = a_2$ then

$$W^{(a|b)} = \alpha_1[(\alpha_2 + \alpha_3)\alpha_4 + \alpha_4(\alpha_5 + \alpha_6 + \alpha_7) + (\alpha_5 + \alpha_6 + \alpha_7)(\alpha_2 + \alpha_3)]$$

THEOREM **3-18** If a, b, c are distinct,

$$[\mathscr{W}^a]^{b,c} = \tilde{W}^{(a|bc)} \tag{3-54}$$

Proof† We first prove

$$[\mathscr{W}^a]^{b,c} + [\mathscr{W}^b]^{a,c} = [\mathscr{W}^a]^{b,b} \tag{3-55}$$

Writing $v(G) = \{a_1, a_2, \ldots, a_M\}$ with $a_1 = a$, $a_2 = b$, and $a_3 = c$, we have

$$[\mathscr{W}^a]^{b,b} = \det(w_{ij}|i = 3, 4, \ldots, M;\ j = 3, 4, \ldots, M)$$

$$[\mathscr{W}^a]^{b,c} = -\det(w_{ij}|i = 3, 4, \ldots, M;\ j = 2, 4, \ldots, M)$$

$$[\mathscr{W}^b]^{a,c} = -\det(w_{ij}|i = 3, 4, \ldots, M;\ j = 1, 4, \ldots, M) \tag{3-56}$$

where

$$w_{ij} \equiv \sum_{l\in|G|} [a_i : l]\,[a_j : l]\,\beta_l \tag{3-57}$$

Hence the first column of $[\mathscr{W}^a]^{b,b} - [\mathscr{W}^a]^{b,c} - [\mathscr{W}^b]^{a,c}$ is $\sum_{j=1}^{3} w_{ij}$, which can be replaced by

$$\sum_{j=1}^{M} w_{ij} = \sum_{l\in|G|} [a_i : l]\,\beta_l \sum_{j=1}^{M} [a_j : l] = 0 \tag{3-58}$$

by adding all the remaining columns to the first one. Thus we have proven

$$[\mathscr{W}^a]^{b,c} + [\mathscr{W}^b]^{a,c} = \tilde{W}^{(a|b)} \tag{3-59}$$

with the aid of (3-48). By permuting a, b, c, we obtain two other equalities. From those three formulas, we find

$$[\mathscr{W}^a]^{b,c} = \tfrac{1}{2}[\tilde{W}^{(a|b)} + \tilde{W}^{(a|c)} - \tilde{W}^{(b|c)}] \tag{3-60}$$

On substituting

$$\tilde{W}^{(a|b)} = \tilde{W}^{(ac|b)} + \tilde{W}^{(a|bc)} \tag{3-61}$$

etc. (see (3-45)), we obtain (3-54). q.e.d.

THEOREM **3-19** The parametric functions U, $W^{(a|b)}$, and W_S at $\alpha_l = 0$ are equal to those in G/l unless they then vanish ($U = 0$ at $\alpha_l = 0$ if l is a

† One can prove this formula by means of the Binet–Cauchy theorem, but to check the sign is rather complicated [9]. We here present a new proof of the author. [Speer [10], however, took it in his book already.]

loop line in G; $W^{(a|b)} = 0$ at $\alpha_l = 0$ if l is a loop line in $G^{(ab)}$; $W_S = 0$ at $\alpha_l = 0$ if either $l \in S$ or l is a loop line in G). The derivatives $\partial U/\partial\alpha_l$, $\partial W^{(a|b)}/\partial\alpha_l$, and $\partial W_S/\partial\alpha_l$ are equal to the U, the $W^{(a|b)}$, and the W_{S-l} in $G - l$, respectively, unless they vanish ($\partial U/\partial\alpha_l = 0$ if l is a cut-line in G; $\partial W^{(a|b)}/\partial\alpha_l = 0$ if l is a cut-line in $G^{(ab)}$; $\partial W_S/\partial\alpha_l = 0$ if l is a cut-line in $G - S$). Dual statements hold for $\tilde{U}$; $\tilde{W}^{(a|b)}$, and $\tilde{W}_S$.

Proof By definition

$$U\big|_{\alpha_l=0} = \sum_{T^* \not\ni l} \prod_{k \in T^*} \alpha_k \tag{3-62}$$

If l is a loop line then (3-62) vanishes because any chord set T^* contains l. Otherwise, it equals the chord-set product sum in G/l because $T^*(\not\ni l)$ is a chord set in G/l [see Theorem 2-7 (b)]. Likewise,

$$\partial U/\partial\alpha_l = \sum_{T^* \ni l} \prod_{k \in T^* - l} \alpha_k \tag{3-63}$$

If l is a cut-line then (3-63) vanishes because any chord set T^* does not contain l. If not so, it equals the chord-set product sum in $G - l$ because $T^* - l$ is a chord set of $G - l$ [see Theorem 2-7(a)]. We can similarly verify the statements for other functions. q.e.d.

THEOREM **3-20** [11] Let U_P be the chord-set product sum in G/P; then

$$\sum_{P \in P(ab)} U_P = U \tag{3-64}$$

where a and b are two arbitrary vertices of G.

Proof Let T^* be an arbitrary chord set in G. Then because of Theorem 2-5(c), its complement T includes a unique path $P \in \boldsymbol{P}(ab)$. Hence $\prod_{l \in T^*} \alpha_l$ is a term of U_P. Conversely, let T^* be a chord set in G/P; then its complement $T = |G| - T^*$ is a tree in G because of Theorem 2-6(b). Hence every term in U_P appears in U. Finally, U_P and $U_{P'}$ for $P \neq P'$ have no common terms because $P \cup P'$ includes a circuit. q.e.d.

§ 4 PLANAR GRAPHS

4-1 Characterizations of the Planar Graph

A *planar graph* is defined to be a graph which can be embedded topologically in a plane, that is, it can be drawn geometrically on a plane without crossing of lines. In general, a planar graph can be realized on a plane in several different ways.

A Feynman graph G is called *planar* if G' is planar, where G' is the graph which is obtained from G by introducing a new vertex "infinity" with which all external lines of G are incident.

For example, Fig. 4-1 (a) is a planar graph because it can be redrawn in a form with no crossing, but the Feynman graphs Fig. 4-1 (b) and (c) are nonplanar.

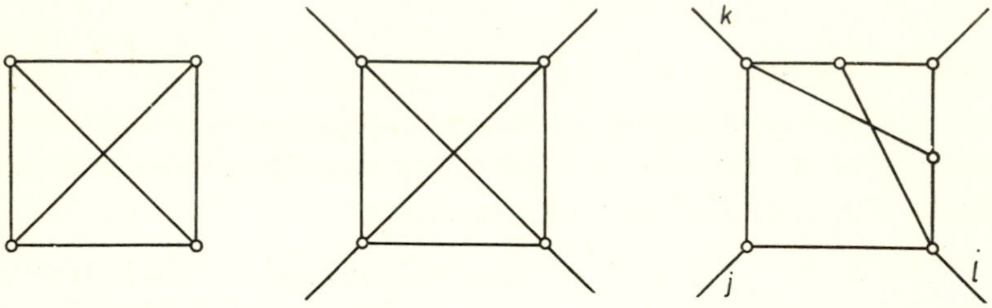

Fig. 4-1 (a) A planar graph. (b) & (c) Nonplanar Feynman graphs.

THEOREM **4-1** A planar graph G has $\mu(=\mu(G))$ independent circuits $C_1, \ldots, C_\mu$ such that any line of G does not belong to more than two of $C_1, \ldots, C_\mu$.

It is easy to see that the boundaries, $C_1, \ldots, C_\mu$, of μ finite, disjoint planar domains are independent. It is convenient to define the overall signs of $[C_j : l]$ $(j = 0, 1, \ldots, \mu)$ in such a way that $[C_i : k] = -[C_j : k]$ for any $k \in C_i \cap C_j$ $(i \neq j)$, where $C_0 \equiv C_1 \oplus \ldots \oplus C_\mu$ is a circuit if G is nonseparable.

It is evident that any subgraph and any reduced graph of a planar graph are planar. The planar graph is characterized by the following celebrated theorem due to Kuratowski [12]. We quote it without proof.

THEOREM **4-2** A necessary and sufficient condition for a graph G to be nonplanar is that there exists a subgraph H of G/I for some $I \subset |G|$ which is one of the Kuratowski graphs shown in Fig. 4-2.

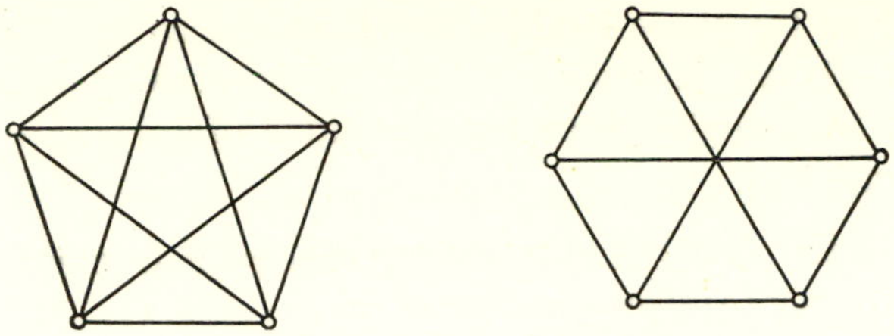

Fig. 4-2 (a) & (b) Kuratowski graphs.

For example, the graph G' corresponding to Fig. 4-1(b) is nothing but Fig. 4-2(a) itself; the graph G' corresponding to Fig. 4-1(c) reduces to Fig. 4-2(b) by considering $(G'/j) - k - l$.

The other, more important characterization of the planar graph is the existence of dual graphs. This will be explained in the next subsection.

4-2 Dual Graphs

Given a planar graph G, when it is drawn on a plane, $\mu(G)$ independent circuits stated in Theorem 4-1 divide that plane into $\mu(G) + 1$ disjoint domains. Then a dual graph $\tilde{G}$ of G is constructed as follows (see Fig. 4-3). We write one vertex of $\tilde{G}$ inside each of those domains and link two vertices

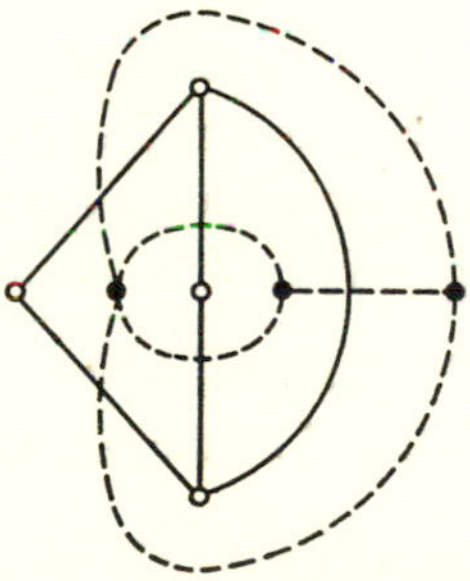

Fig. 4-3 An example of constructing a dual graph. Dotted lines indicate the lines of the dual graph.

of $\tilde{G}$ by r lines of $\tilde{G}$ when the boundaries (i.e., circuits) of the corresponding two domains of G have r common lines of G. In particular, cut-lines in G correspond to loop lines in $\tilde{G}$. Since in general G may have another set of $\mu(G)$ independent circuits satisfying the condition stated in Theorem 4-1, dual graphs of G will not necessarily be unique.

As seen from the above intuitive definition, a dual graph exists if and only if G is planar. In order to state this proposition mathematically, however, we have to give an abstract definition of a dual graph, which is due to Whitney [13].

DEFINITION **4-3** A graph $\tilde{G}$ is called a *dual graph* of G if the following two conditions are satisfied:

(a) $|G|$ and $|\tilde{G}|$ are in one-to-one correspondence.

(b) Let $\tilde{H}$ be a subgraph of $\tilde{G}$ which corresponds to a subgraph H of G by the above correspondence. Then for any H of G we have

$$r(\tilde{H}^*) = r(\tilde{G}) - \mu(H) \tag{4-1}$$

where H^* is a subgraph of G such that $|H^*| = |G| - |H|$.

It is convenient to assume in the above definition that no isolated points are contained in G and $\tilde{G}$ and also in any of their subgraphs considered. From Definition 4-3, the following properties of $\tilde{G}$ follow immediately:

$$N(\tilde{G}) = N(G), \quad N(\tilde{H}) = N(H) \tag{4-2}$$

$$\mu(\tilde{G}) = r(G), \quad r(\tilde{G}) = \mu(G) \tag{4-3}$$

$$\mu(\tilde{H}^*) = \mu(\tilde{G}) - r(H) \tag{4-4}$$

The third property is equivalent to (4-1) because of (2-28).

Definition 4-3 is not intuitive, but it makes sense for any graph. Whitney [14] proved, however, that it coincides with the intuitive definition. We here quote his theorem without proof.

THEOREM **4-4** A graph G has a dual graph $\tilde{G}$ if and only if G is planar, and $\tilde{G}$ can be constructed as stated in the beginning of this subsection.

The incidence number $[\tilde{a}_j, \tilde{l}]$ of $\tilde{G}$ is defined by $[C_j : l]$ of G if $\tilde{a}_j$ and $\tilde{l}$ correspond to C_j and l, respectively. Likewise, the circuit numbers $[\tilde{C}, \tilde{l}]$ for $M - 1$ independent circuits of $\tilde{G}$ are defined by the corresponding incidence numbers of G.

The following duality properties can easily be verified.

THEOREM **4-5** If $\tilde{G}$ is a dual graph of G, G is a dual graph of $\tilde{G}$.

THEOREM **4-6** If G is nonseparable then $\tilde{G}$ is also nonseparable. [However, if G is connected then $\tilde{G}$ is not necessarily connected.]

THEOREM **4-7** If $\tilde{G}$ is a dual graph of G, this "dual" correspondence between G and $\tilde{G}$ takes each concept listed in Subsection 2-3 into its dual concept. Especially, a circuit and a cut-set in G uniquely corresponds to a cut-set and a circuit in $\tilde{G}$, respectively.

Thus the "dual" correspondence respects the duality stated in Subsection 2-3. The latter makes sense for any graph, but a dual graph exists only for a planar graph. The reason for this is that the fundamental concept, vertex, of a graph has no dual except for a planar graph, in which a domain on a plane is the dual concept of a vertex. Thus the existence of the vertex plays a role of a "symmetry breaking" of the duality.

Next, we are concerned with the nonuniqueness of dual graphs. As mentioned above, dual graphs of a graph G are not necessarily unique, but they bear some resemblance to each other.

DEFINITION **4-8** Two graphs G_1 and G_2 are called *2-isomorphic* if G_2 is obtained from G_1 through the following (reversible) operations:

(a) To divide a connected component into two parts at a cut-vertex, and its reversed operation.

(b) To interchange the incidence with a of the lines of $|H| \cap S[a]$ and that with b of the lines of $|H| \cap S[b]$ when there are two subgraphs H and H^* such that $|H^*| = |G| - |H|$ and $v(H) \cap v(H^*) = \{a, b\}$. (See Fig. 4-4).

THEOREM **4-9** If and only if G_1 and G_2 are 2-isomorphic, there is a one-to-one correspondence of $|G_1|$ and $|G_2|$ which gives a one-to-one correspondence of circuits of G_1 and G_2.

We omit the proof of this theorem [15]. The following theorem follows directly from Theorems 4-7 and 4-9.

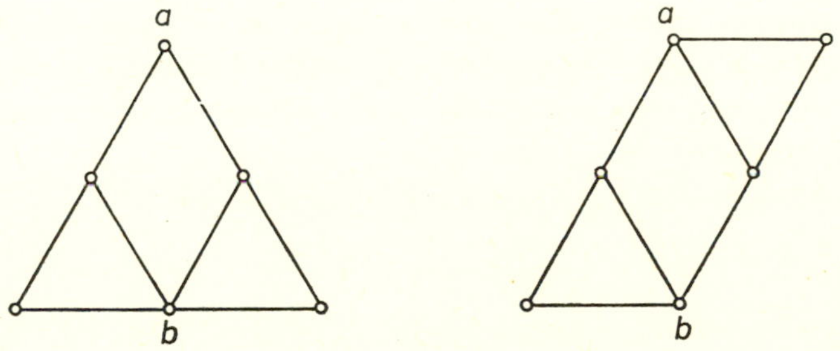

Fig. 4-4 2-isomorphic graphs.

THEOREM 4-10 Dual graphs of a graph G are mutually 2-isomorphic.

For example, the graph shown in Fig. 4-5 has two different dual graphs shown in Fig. 4-4.

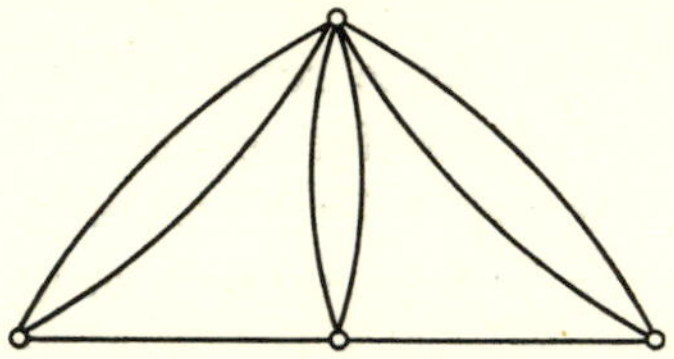

Fig. 4-5 A graphs which has two topologically different dual graphs shown in Fig. 4-4.

Finally, for a Feynman graph G, its dual graph $\tilde{G}$ is defined as a dual graph of G' (see Subsection 4-1). It is customary to draw all the external lines of G in the outside of all the circuits of G when G' is drawn on a plane. Hence the outermost circuit of $\tilde{G}$ is formed only by the ones corresponding to the external lines of G because the outside domain of $\tilde{G}$ corresponds to the point "infinity" of G'.

§5 TRANSPORT PROBLEM

In this section, we deal with a rather special problem, which occurs in the theory of linear programming. We consider a graph having n special vertices, which will be identified with the n external vertices in a Feynman graph. The problem is to transport given loads placed at some of those vertices to the remainders as requested in such a way that when carrying a load along a line l it does not exceed the capacity assigned to l. We call this problem the *transport problem*, which will be applied to the theory of the Feynman integral in order to find some support properties (see Section 19).

To be more precise, we first present some definitions. We consider a connected graph G, and let g ($\subset v(G)$) be a set of n special vertices. For each line l of G, we assign a real, nonnegative quantity $c(l)$, which is called a *capacity*. For each $a \in g$, we assign a real quantity, called a *demand*, $d(a)$, in such a way that the conversation law

$$\sum_{a \in g} d(a) = 0 \tag{5-1}$$

is satisfied. A *flow* $f(l)$ is a function of $l \in |G|$ such that

$$\sum_{l \in |G|} [a : l] f(l) = 0 \quad \text{for any } a \in v(G) - g \tag{5-2}$$

$$|f(l)| \leqq c(l) \quad \text{for any } l \in |G| \tag{5-3}$$

Then the problem is to find under what conditions a flow satisfying

$$\sum_{l \in |G|} [a : l] f(l) = d(a) \quad \text{for any } a \in g \tag{5-4}$$

exists. We note that (5-1) is a necessary consequence of (5-4) and (5-2).

To solve this problem, it is convenient to introduce an auxiliary graph $\bar{G}$ such that $v \equiv v(\bar{G}) = v(G)$ and $\bar{G}$ is a complete graph. Since any line of $\bar{G}$ is in one-to-one correspondence to a pair of two distinct vertices, we define a capacity and a flow in $\bar{G}$ in the following way. A capacity $c(a,b) \equiv c(b,a)$ is a function of an unordered pair of vertices a and b, which is defined by

$$c(a, b) \equiv \sum_{l \in S[a] \cap S[b]} c(l) \tag{5-5}$$

for $a \neq b$ and $c(a, a) \equiv 0$. Likewise, a flow $f(a, b)$ is a function of an ordered pair of a and b, which is defined by

$$f(a, b) \equiv \sum_{l \in S[b]} [a : l] f(l) = - \sum_{l \in S[a]} [b : l] f(l) \tag{5-6}$$

for $a \neq b$, and $f(a, a) \equiv 0$. From the above definitions together with (5-2) and (5-3), we obtain

$$f(a, b) = -f(b, a) \quad \text{for any } a, b \in v \tag{5-7}$$

$$\sum_{b \in v} f(a, b) = 0 \quad \text{for any } a \in v - g \tag{5-8}$$

$$|f(a, b)| \leqq c(a, b) \quad \text{for any } a, b \in v \tag{5-9}$$

The demand $d(a)$ in $\bar{G}$ is defined to be the same as that in G. Hence (5-4) becomes

$$\sum_{b \in v} f(a, b) = d(a) \quad \text{for any } a \in g \tag{5-10}$$

Then it is straightforward to show that the transport problem in G is equivalent to that in $\bar{G}$. [Note that given $f(a, b)$, $[a : l]f(l)$ for $l \in S[a] \cap S[b]$ should be chosen to have the same sign.]

In order to simplify the description, we abbreviate a sum of $c(a, b)$, $f(a, b)$ or $d(a)$ over a set of vertices by writing the name of the set at the place of the vertex, e.g.,

$$f(u, b) \equiv \sum_{a \in u} f(a, b) \tag{5-11}$$

Then we have additive functions over v. The conditions (5-7), (5-8) and (5-9) are rewritten as

$$f(u, u) = 0 \quad \text{for any } u \subset v \tag{5-12}$$

$$f(u, v) = 0 \quad \text{for any } u \subset v - g \tag{5-13}$$

$$|f(u, u')| \leqq c(u, u') \quad \text{for any } u, u' \subset v \tag{5-14}$$

respectively. Since $f(v - g, v) = 0$ and

$$f(g, v) + f(v - g, v) = f(v, v) = 0 \tag{5-15}$$

from (5-13) and (5-12), respectively, we obtain

$$f(g, v) = 0 \tag{5-16}$$

which represents an overall conservation law.

We first consider the case $n = 2$, in which $g = \{a, a'\}$. Then (5-16) reduces to

$$f(a, v) = -f(a', v) \tag{5-17}$$

We may assume $f(a, v) \geqq 0$ without loss of generality. The following *minimum cut theorem* is due to Ford and Fulkerson [16].

THEOREM **5-1** The maximum value of $f(a, v)$ is equal to the minimum of the capacities of cut-sets belonging to $S_{\bar{G}}(a \mid a')$, that is,

$$\max_f f(a, v) = \min_{u(\ni a, \not\ni a')} c(u, v - u) \tag{5-18}$$

Proof [17] We first observe that for any flow f

$$f(a, v) \leqq c(u, v - u) \quad \text{for any } u(\ni a, \not\ni a') \tag{5-19}$$

because

$$0 \leqq f(a, v) = f(u, v) = f(u, u) + f(u, v - u)$$
$$= f(u, v - u) \leqq c(u, v - u) \tag{5-20}$$

with the aid of (5-13), (5-12), and (5-14). Hence we have only to prove that there exists $u \subset v$ such that $a \in u$, $a' \notin u$, and

$$\max_f f(a, v) = c(u, v - u) \tag{5-21}$$

We denote by $\bar{f}$ one of the flows which realize the maximum of $f(a, v)$. [From the definition, it is evident that there exists $\bar{f}$ such that $\bar{f}(a, v) = \sup_f f(a, v)$.]

An *a-path* is a sequence of vertices $P \equiv (a_0, a_1, \ldots, a_m)$ such that its first vertex is $a_0 = a$ and

$$r_j \equiv c(a_{j-1}, a_j) - \bar{f}(a_{j-1}, a_j) > 0 \quad (j = 1, 2, \ldots, m) \tag{5-22}$$

Let u be the set consisting of a and all the vertices each of which is the end point a_m of some a-path. Then we show $a' \notin u$ as follows. If $a' \in u$ then there would exist an a-path $P = (a_0, a_1, \ldots, a_m = a')$ so that $r \equiv \min_{1 \leqq j \leqq m} r_j > 0$, whence there would exist a flow $\bar{f}'$ such that $\bar{f}'(a, v) = \bar{f}(a, v) + r$ in contradiction to the maximality of $\bar{f}(a, v)$, where $\bar{f}'$ is defined by

$$\bar{f}'(a_{j-1}, a_j) \equiv \bar{f}(a_{j-1}, a_j) + r \quad (j = 1, 2, \ldots, m) \tag{5-23}$$

and $\bar{f}'(b, b') \equiv \bar{f}(b, b')$ for any other pair (b, b').

Finally, we show

$$\bar{f}(a, v) = c(u, v - u) \tag{5-24}$$

If not, (5-20) shows

$$\bar{f}(a, v) = \bar{f}(u, v - u) < c(u, v - u) \tag{5-25}$$

Then there have to exist $b \in u$ and $b' \in v - u$ such that

$$c(b, b') - \bar{f}(b, b') > 0 \tag{5-26}$$

Since $b \in u$, however, there exists an a-path $P = (a, a_1, \ldots, b)$, and it is extendable to $P' = (a, a_1, \ldots, b, b')$ because of (5-26). The existence of an a-path P' is contradictory to $b' \notin u$. Thus (5-24), i.e., (5-21) has to hold. q.e.d.

Now, we return to the general case. The problem is to find a flow satisfying (5-10), namely

$$f(a, v) = d(a) \quad \text{for any } a \in g \tag{5-27}$$

The following theorem is due to Gale [17], but it is slightly modified because we have imposed on a flow the exact conservation law (5-8) at each vertex.

THEOREM **5-2** A necessary and sufficient condition for the existence of a flow satisfying (5-27) is

$$c(u, v - u) \geqq |d(u \cap g)| \quad \text{for any } u \subset v \tag{5-28}$$

Proof It is convenient to define

$$d(b) = 0 \quad \text{for any } b \in v - g \tag{5-29}$$

Then (5-1) and (5-28) are rewritten as

$$d(v) = 0 \tag{5-30}$$

$$c(u, v - u) \geqq |d(u)| \quad \text{for any } u \subset v \tag{5-31}$$

respectively. Furthermore, (5-13) and (5-27) are combined into

$$f(u, v) = d(u) \quad \text{for any } u \subset v \tag{5-32}$$

If f satisfies (5-12), (5-14), and (5-32), then

$$|d(u)| = |f(u, v)| = |f(u, v - u)| \leqq c(u, v - u) \tag{5-33}$$

that is, we have (5-31). Therefore, we have only to prove the converse, that is, we show that if d satisfies (5-31) together with (5-30) then we can find f satisfying (5-12), (5-14), and (5-32).

We introduce a new complete graph $\bar{G}'$ by adding two vertices a and a' to $\bar{G}$, that is, $v(\bar{G}') = v + a + a' \equiv v'$. Let u_0 be the set of all vertices $b \in v$ such that $d(b) \geqq 0$. The capacity c' in $\bar{G}'$ is defined by

$$\begin{aligned} c'(b, b') &= c(b, b') && \text{for any } b, b' \in v \\ c'(a, b) &= d(b) && \text{for any } b \in u_0 \\ c'(b, a') &= -\,d(b) && \text{for any } b \in v - u_0 \\ c'(b, b') &= 0 && \text{otherwise} \end{aligned} \tag{5-34}$$

We first show that

$$c'(a, v' - a) = c'(v' - a', a') = \min_{u'(\ni a, \not\ni a')} c'(u', v' - u') \tag{5-35}$$

in $\bar{G}'$. Let $u \equiv u' - a \subset v$. From (5-34) we have

$$\begin{aligned} c'(u', v' - u') &= c'(u, v - u) + c'(a, v - u) + c'(u, a') + c'(a, a') \\ &= c(u, v - u) + d((v - u) \cap u_0) - d(u \cap (v - u_0)) \end{aligned} \tag{5-36}$$

$$c'(a, v' - a) = d(u_0) = d((v - u) \cap u_0) + d(u \cap u_0) \tag{5-37}$$

By subtracting (5-36) from (5-37), we find

$$\begin{aligned} &c'(a, v' - a) - c'(u', v' - u') \\ &\quad = d(u \cap u_0) + d(u \cap (v - u_0)) - c(u, v - u) \\ &\quad = d(u) - c(u, v - u) \leqq 0 \end{aligned} \tag{5-38}$$

because of (5-31). Thus $c'(a, v' - a)$ is the minimum of $c'(u', v' - u')$. On putting $u' = v' - a'$, (5-38) reduces to

$$c'(a, v' - a) - c'(v' - a', a') = d(v) = 0 \tag{5-39}$$

Thus (5-35) is established.

Now, from Theorem 5-1, there exists a flow f' for $g = \{a, a'\}$ such that

$$f'(a, v' - a) = c'(a, v' - a) = c'(v' - a', a') = f'(v' - a', a') \tag{5-40}$$

together with

$$f'(u, v') = 0 \quad \text{for any } u \subset v \tag{5-41}$$

With the aid of (5-14) for f' and c', (5-40) implies

$$\begin{aligned} f'(a, u) &= c'(a, u) = d(u \cap u_0) \\ f'(u, a') &= c'(u, a') = -d(u \cap (v - u_0)) \text{ for any } u \subset v \end{aligned} \tag{5-42}$$

Let $f(b, b') \equiv f'(b, b')$ for any $b, b' \subset v$; then f evidently satisfies (5-12) and (5-14). From (5-41) and (5-42) we find

$$\begin{aligned} f(u, v) &= f'(u, v') - f'(u, a) - f'(u, a') \\ &= f'(a, u) - f'(u, a') \\ &= d(u \cap u_0) + d(u \cap (v - u_0)) \\ &= d(u) \quad \text{for any } u \subset v \end{aligned} \tag{5-43}$$

Thus (5-32) is satisfied. q.e.d.

Now, we translate the above theorem into a theorem in the original graph G.

THEOREM **5-3** A necessary and sufficient condition for the existence of a flow satisfying (5-4) is

$$\sum_{l \in S} c(l) \geqq |d(h)| \quad \text{for any } h \subset g \text{ and any } S \in \mathbf{S}(h|g-h) \tag{5-44}$$

Proof In G, (5-28) becomes

$$\sum_{l \in \bar{S}} c(l) \geqq |d(u \cap g)| \quad \text{for any } u \subset v \tag{5-45}$$

where $\bar{S}$ is the cut in G dividing v into u and $u^* = v - u$. On the other hand, on putting $h = u \cap g$, (5-44) can be rewritten as

$$\sum_{l \in S} c(l) \geqq |d(u \cap g)| \quad \text{for any } \mathbf{S}(u|u^*) = \{S\} \neq \emptyset \tag{5-46}$$

Therefore, we have only to show the equivalence between (5-45) and (5-46). Since a cut-set is a cut, (5-46) follows from (5-45). Hence we prove the converse. As discussed in the proof of Theorem 2-4(b), the cut $\bar{S}$ is a disjoint union of cuts $\bar{S}_1, \ldots, \bar{S}_k$, which are stars in $G/\bar{S}^*$. By construction, $\bar{S}_j$ divides v into u_j and $v - u_j$ in such a way that u is a disjoint union of $u_1, \ldots, u_k$. [Note that u_j is a vertex in $G/\bar{S}^*$.] Furthermore, each $\bar{S}_j$ is a disjoint union of cut-sets $S_{ji} \in \mathbf{S}(u_{ji} \mid u_{ji}^*)$ such that $v - u_j$ is a disjoint union of u_{ji} over i. Because of the triangular inequality, we have

$$\sum_{i} |d(u_{ji} \cap g) \geqq |d((v - u_j) \cap g)| = |d(u_j \cap g)| \tag{5-47}$$

$$\sum_{j=1}^{k} |d(u_j \cap g)| \geqq |d(u \cap g)| \tag{5-48}$$

Hence (5-46) leads us to

$$\sum_{l \in \bar{S}} c(l) = \sum_{j=1}^{k} \sum_{i} \sum_{l \in S_{ji}} c(l) \geqq \sum_{j=1}^{k} \sum_{i} |d(u_{ji} \cap g)|$$
$$\geqq |d(u \cap g)| \tag{5-49}$$

q.e.d.

In the above theorem, if $c(l)$ and $d(a)$ are restricted to integer values, $f(l)$ is also integer-valued, because all the above consideration can be made also by restricting them to integers. In particular, choosing $c(l) = 1$ for all $l \in |G|$, we obtain the following *path theorem* [18].†

THEOREM **5-4** Let $g \subset v$ and $h_0 \subset g$. If

$$N(S) \geqq \left| \sum_{a \in h \cap h_0} r_a - \sum_{a \in h \cap (g - h_0)} r_a \right|$$

$$\text{for any } h \subset g \text{ and any } S \in \mathcal{S}(h|g-h) \tag{5-50}$$

where positive integers r_a $(a \in g)$ satisfy

$$r \equiv \sum_{a \in h_0} r_a = \sum_{a \in g - h_0} r_a \tag{5-51}$$

then we can find r disjoint paths such that for any $a \in h_0$ there exist r_a paths which have a as their end point and such that for every path one end point belongs to h_0 and the other belongs to $g - h_0$.

References

(1) S. Seshu and M. B. Reed, *Linear Graphs and Electrical Networks* (Addison Wesley Mass., 1961).
(2) C. Berge, *Théorie des Graphs et Ses Applications*, 2nd ed. (Dunod, Paris, 1963); English Translation by A. Doig (John Wiley & Sons, New York).
(3) O. Ore, *Theory of Graphs* (American Math. Soc., 1962).
(4) F. Harary (editor), *Graph Theory and Theoretical Physics* (Academic Press, New York, 1967).
(5) R. P. Feynman, *Phys. Rev.* **76**, 749 (1949).
(6) R. P. Feynman, *Phys. Rev.* **76**, 769 (1949).
(7) H. W. Turnbull, *The Theory of Determinants, Matrices, and Invariants*, 3rd ed. (Dover, New York, 1960), p. 82.
(8) N. Nakanishi, *Progr. Theoret. Phys.* **17**, 401 (1957).
(9) Ref. 1, p. 161.
(10) E. R. Speer, *Generalized Feynman Amplitudes* (Princeton University Press, Princeton, N. J., 1969).
(11) Y. Chow and D. J. Kleitman, *Progr. Theoret. Phys.* **32**, 950 (1964).
(12) C. Kuratowski, *Fund. Math.* **15**, 271 (1930).
(13) Ref. 1, p. 41.
(14) Ref. 1, p. 45.
(15) H. Whitney, *Am. J. Math.* **55**, 245 (1933).
(16) L. R. Ford Jr. and D. R. Fulkerson, *Canad. J. Math.* **8**, 399 (1956).
(17) D. Gale, *Pacific J. Math.* **7**, 1073 (1957).
(18) N. Nakanishi, *Progr. Theoret. Phys. Suppl.* **18**, 1 (1961); Errata, *Progr. Theoret. Phys.* **26**, 806 (1961).

† Nakanishi derived Theorem 5-3 from Theorem 5-4 by taking the limit of an infinite number of lines.

CHAPTER 2

Feynman-Parametric Formula

The best known theory for describing reactions of elementary particles is quantum field theory. According to it, the probability amplitudes of reactions are conveniently described by an S-matrix, which may be formally written in terms of a power series of interaction strengths. Each term of it is a sum of the Feynman integrals corresponding to the Feynman graphs specified by the type of reaction, the types of interactions assumed, and the order of the term.

A Feynman integral is a multiple, singular integral over infinite ranges of 4-momenta, and it is not necessarily well defined *a priori*. It is important, therefore, from both theoretical and practical points of view, to convert the Feynman integral into a Feynman-parametric integral whose integration region is compact. The latter contains certain functions of Feynman parameters in the integrand, and they can be expressed in terms of some topological formulas presented in Subsection 3-2, which directly reflect the topological structure of the Feynman graph.

In 1949, Feynman [1] introduced the *Feynman identity*

$$\frac{1}{AB} = \int_0^1 \frac{d\alpha}{[\alpha A + (1 - \alpha)B]^2}$$

where α is called a Feynman parameter, in order to calculate some low-order Feynman integrals. In 1951, Chisholm [2] obtained a Feynman-parametric formula for the general Feynman integral by means of repeated use of the above Feynman identity. In 1956, Nakanishi [3] employed a generalized Feynman identity in order to deal with all Feynman parameters homogeneously, and presented topological formulas for various parametric functions. On the other hand, on the basis of Nambu's inverse-Feynman-parametric representation [4], in 1958 Symanzik [5] presented a convenient expression for the most important parametric function, called the V function, without proof. Three independent proofs of the Nambu-Symanzik formula for V from its expression presented by Nakanishi were given by Nakanishi [6], by Shimamoto [7], and in a slightly modified form by Kinoshita [8]. In his derivation, Shimamoto first introduced the graph-theoretical approach to the theory of Feynman integrals.

In Section 6, we first review the physical background of the Feynman integral for the sake of nonphysicist's convenience, and then give the formal definition of the Feynman integral. In Section 7, we convert it into a Feynman-parametric integral and present various topological formulas for the important parametric function V. Section 8 is devoted to the position-space approaches to the Feynman integral. In Section 9, we discuss how the Feynman-parametric integral is modified when nonzero-spin particles are present. Finally, we present a rigorous proof of Dyson's power-counting theorem for ultraviolet divergence and briefly describe the renormalization of the Feynman-parametric integral.

§ 6 FEYNMAN INTEGRALS

6-1 Physical Background

Nowadays, many elementary particles are known to exist. They are classified into four families: *baryons*, *mesons*, *leptons*, and *photons*. Baryons and leptons are *fermions* having spin $\frac{1}{2}$, while mesons and photons are *bosons* having spin 0 or 1,† where *spin* is an intrinsic angular momentum of a particle in natural units.‡ A proton and a neutron are examples of baryons. They are altogether called *nucleons*, while the baryons other than nucleons are called *hyperons*, which are heavier than nucleons. An electron is an example of a lepton. The numbers of baryons and of leptons are known to be strictly conserved (at least within the ordinary time scale) separately if the numbers of their antiparticles are counted as minus, where an *antiparticle* is a particle having "reversed" quantum numbers.

Elementary particles interact each other and can be converted into other particles if various conservation laws such as energy-momentum conservation, baryon number conservation, etc. are satisfied. Apart from the gravitational interaction, there are three kinds of interactions, called *strong*, *electromagnetic*, and *weak*. The strong and the weak interactions are stronger than and much weaker than the electromagnetic interaction, respectively, in their strengths. Only baryons and mesons interact strongly, and therefore they are altogether called *hadrons*.

The transition probability of any reaction is given by an absolute square of the corresponding probability amplitude. The probability amplitudes between all possible pairs of states constitutes an *S-matrix*, where states are specified by the numbers and the kinds of particles, their energies and momenta, and other quantum numbers. As explained below, the S-matrix can be given explicitly, at least formally, on the basis of the quantum field theory, where the *quantum field theory* is a theory describing fields which have the ability of creating and annihilating the corresponding particles.

Let $x = (x^{(0)}, x^{(1)}, x^{(2)}, x^{(3)})$ be coordinates of a point in the four-dimensional space-time, which is called the *position space*; $x^{(0)}$ stands for the

† There are also many resonances (including an elementary particle Ω) which may have higher spins.

‡ The natural unit system is defined by taking $c = h = 1$, where c is the speed of the light and $2\pi h$ is the Planck's constant (the quantum of action). Hereafter we always work in natural units.

time coordinate and $(x^{(1)}, x^{(2)}, x^{(3)})$ represents a point in the ordinary three-dimensional euclidean space. Let a be an arbitrary 4-vector $(a^{(0)}, a^{(1)}, a^{(2)}, a^{(3)})$; then a transformation from x to $x + a$ for all x is called a *translation*. Let Λ be an arbitrary, linear, homogeneous transformation of x; it can be represented by a 4×4 matrix. If Λ leaves a quadratic form $x^2 \equiv xx$ invariant, where

$$xy \equiv x^{(0)}y^{(0)} - x^{(1)}y^{(1)} - x^{(2)}y^{(2)} - x^{(3)}y^{(3)} \tag{6-1}$$

with $y = (y^{(0)}, y^{(1)}, y^{(2)}, y^{(3)})$, then it is called a *Lorentz transformation*. The totality of the Lorentz transformations forms a group, called the *Lorentz group*, which consists of four connected parts. Elements of the maximal connected subgroup of the Lorentz group are called *Lorentz rotations*. The group generated by all translations and Lorentz transformations is called the *Poincaré group*.

The quantum field theory should be invariant under the Poincaré group, or more precisely, under all translations and Lorentz rotations. A *field operator*, or simply a *field*, is an operator-valued distribution of x which has a certain correspondence with a particle. Since the field $\psi(x)$ should be invariant under translations because of the homogeneity of our space, $\psi(x)$ should transform covariantly under the Lorentz group, that is, the field forms a representation of the Lorentz group. The spin of a particle characterizes the representation of the corresponding field, that is, the former is an eigenvalue of a Casimir operator. Particles having spin 0 are simply called *spinless* particles.

If a particle exists solely, that is, if any other particles are at very large distances from it, then it is called *free*. The wave function $\varphi(x)$ of a free particle should satisfy the *Klein–Gordon equation*

$$[m^2 + (\partial/\partial x)^2]\varphi(x) = 0 \tag{6-2}$$

where a *wave function* is a complex-valued function of x representing a (quantum-mechanical) state and m stands for the mass of the particle, which is a real non-negative constant for any known particle. If we consider a particle having nonzero spin, $\varphi(x)$ consists of several components. For example, if we consider a particle having spin $\frac{1}{2}$, which is called a *Dirac particle*, $\varphi(x)$ consists of four components and satisfies the *Dirac equation*

$$(m - i\gamma\partial/\partial x)\varphi(x) = 0 \tag{6-3}$$

where $\gamma = (\gamma^{(0)}, \gamma^{(1)}, \gamma^{(2)}, \gamma^{(3)})$ and $\gamma^{(0)}$ and $\gamma^{(j)}$ $(j = 1, 2, 3)$ are 4×4 constant matrices satisfying

$$\gamma^{(\mu)}\gamma^{(\nu)} + \gamma^{(\nu)}\gamma^{(\mu)} = 2\delta^{(\mu\nu)} \quad (\mu, \nu = 0, 1, 2, 3) \tag{6-4}$$

where $\delta^{(\mu\nu)}$ is a Minkowski metric tensor defined by $\delta^{(00)} = -\delta^{(jj)} = 1$ and $\delta^{(\mu\nu)} = 0$ for $\mu \neq \nu$. It is easy to verify that any solution to a Dirac equation (6-3) satisfies (6-2). A Dirac particle with $m \neq 0$ necessarily has its anti-particle; two of the four components of $\varphi(x)$ correspond to the freedoms of the latter.

It is convenient to consider a free field operator $\psi(x)$ though it is somewhat unphysical; it satisfies the same equation as that of $\varphi(x)$, but it is an operator which can create or annihilate the corresponding particle in any free state. For simplicity, let us consider a spinless particle which has no distinct anti-particle. A complete set of solutions to (6-2) is given by e^{-ipx} and e^{ipx}, where p is a 4-momentum $(p^{(0)}, p^{(1)}, p^{(2)}, p^{(3)})$ with the energy

$$p^{(0)} \equiv [m^2 + (p^{(1)})^2 + (p^{(2)})^2 + (p^{(3)})^2]^{\frac{1}{2}} \geqq 0 \tag{6-5}$$

Hence $\psi(x)$, which is hermitian in our case, can be expanded into

$$\psi(x) = \prod_{j=1}^{3} \left(\int_{-\infty}^{\infty} dp^{(j)} \right) [a(p)e^{-ipx} + a^\dagger(p)e^{ipx}] \tag{6-6}$$

where the dagger denotes hermitian conjugation. The operators $a(p)$ and $a^\dagger(p)$ represent an annihilation and a creation operator of a particle having a 4-momentum p, respectively. They satisfy the *commutation relations*

$$[a(p), a(p')] = 0$$

$$[a(p), a^\dagger(p')] = (2p^{(0)})^{-1} \prod_{j=1}^{3} \delta(p^{(j)} - p'^{(j)}) \tag{6-7}$$

where $[A, B] \equiv AB - BA$. The coefficient $(2p^{(0)})^{-1}$ has been introduced owing to the requirement of the Lorentz invariance.

The operand of $a(p)$ and $a^\dagger(p)$ is called a (second-quantized) *state*, which differs from the quantum-mechanical state because the former contains information on the numbers of particles. The state in which no particles are present is called the *vacuum*, and denoted by $|0\rangle$. It is a Lorentz-invariant state, and characterized by

$$a(p)|0\rangle = 0 \quad \text{for any } p \tag{6-8}$$

The adjoint of a state $|\alpha\rangle$ is denoted by $\langle\alpha|$ so that $\langle\beta\,|\,\alpha\rangle$ represents an inner product of $|\alpha\rangle$ and $|\beta\rangle$; the totality of the states spans a Hilbert space.†

From (6-6) and (6-7), we obtain

$$[\varphi(x), \varphi(y)] = \int d^4k\,(\mathrm{sgn}\,k^{(0)})\,\delta\,(m^2 - k^2)e^{-ik(x-y)} \tag{6-9}$$

where $\int d^4k$ stands for the integration over all 4-momentum space of $k = (k^{(0)}, k^{(1)}, k^{(2)}, k^{(3)})$, and $\mathrm{sgn}\,k^{(0)} = k^{(0)}/|k^{(0)}|$. [For the meaning of the δ-function used here, see Appendix A-1.] Since the right-hand side of (6-9) vanishes for $x^{(0)} = y^{(0)}$ with $x \neq y$ and is Lorentz invariant, we have

$$[\varphi(x), \varphi(y)] = 0 \quad \text{for } (x-y)^2 < 0 \tag{6-10}$$

Such a relation as (6-10) is called the *microcausality condition.* It is closely related to Einstein's relativity principle, which states that any action cannot propagate faster than the light speed ($c = 1$).

Now, we consider field operators which are not free. Since various elementary particles interact each other with some regularity, we have to specify in what ways various field operators are mutually related; it is called *dynamics.* To specify dynamics, it is the orthodox way to consider a Lagrangian L of the system; L is a Lorentz invariant, hermitian, operator-valued functional of various fields. It consists of a *free Lagrangian* L_{free} and and *interaction Lagrangian* L_{int}. The Euler equations yielded by L_{free} are equations for free fields, while L_{int} introduces additional terms to them; for example, the Euler equation yielded by L for a spinless particle is

$$[m^2 + (\partial/\partial x)^2]\,\boldsymbol{\varphi}(x) = \boldsymbol{\varrho}(x) \tag{6-11}$$

where $\boldsymbol{\varrho}(x)$ is nonlinear with respect to fields and represents interactions. The field operator which satisfies an equation containing an interaction term such as (6-11) is called a *Heisenberg operator.* It is prohibitive to describe its operator properties completely, but it is usually supposed to satisfy the microcausality condition such as

$$[\boldsymbol{\varphi}(x), \boldsymbol{\varphi}(y)] = 0 \quad \text{for } (x-y)^2 < 0 \tag{6-12}$$

Because of the translational invariance of the theory, there exist four generators of translations, which are denoted by $P = (P^{(0)}, P^{(1)}, P^{(2)}, P^{(3)})$.

† The above description of operators and states is not mathematically rigorous. We here intend to explain the theory intuitively.

Their simultaneous eigenstates form a complete set of states. The eigenvalues of $P^{(0)}$ and P^2 are supposed to be real and nonnegative because they represent the energy and the invariant square of the 4-momentum of the system, respectively. From physical consideration, the existence region of the eigenvalues of P^2 is usually further restricted. This restriction is called a *spectral condition*. The *axiomatic field theory* [9] is a theory which starts from postulates such as the invariance under the Poincaré group, the Hilbert-space properties of states, and microcausality and spectral conditions. In the axiomatic field theory, neither the existence of a Lagrangian nor detailed operator properties of fields are assumed. Therefore this theory cannot describe the system completely, but from it we can obtain certain amounts of information about the general framework of the quantum field theory. For example, with one additional postulate, one can show that the S-matrix exists and is unitary for the physical values of 4-momenta (this fact represents the conservation of total probability).

In order to find the S-matrix explicitly, we have to specify not only the Lagrangian but also the operator properties of fields completely. To do this, it is convenient to make a (formal) unitary transformation of states in such a way that equations of fields reduce to those of the corresponding free fields. The transformed states are called the states in the *interaction representation*, which are time-dependent. We can describe those states Lorentz-covariantly by using a certain device, but we here omit details. In the interaction representation, we know the commutation relations of fields completely such as (6-9), and therefore the theory is fully specified. Given a state at a particular time, we can in principle calculate a state in any time by means of the Hamiltonian formalism. The state at $x^{(0)} \to -\infty$ and that at $x^{(0)} \to +\infty$ are called the *initial* and the *final* state, respectively. The particles in those states are supposed to be free. The inner product of the initial state and the final one is nothing but the S-matrix element. After a lot of analysis [10], which is known as the *perturbation theory*, one finds that the S-matrix is given by $S \equiv \tilde{S}/\langle 0\,|\tilde{S}|\,0\rangle$, where

$$\tilde{S} \equiv \sum_{M=0}^{\infty} \frac{i^M}{M!} \prod_{j=1}^{M} \left(\int d^4 x_j \right) T[\mathscr{L}_{\text{int}}(x_1) \ldots \mathscr{L}_{\text{int}}(x_M)] \tag{6-13}$$

Here

$$L_{\text{int}} = \int d^4 x \, \mathscr{L}_{\text{int}}(x) \tag{6-14}$$

and the symbol T indicates that the succeeding operators should be arranged in the order of times $x_j^{(0)}$. The series expansion of S, which is called a *perturbation series*, is only formal because, as we see later, it is not well-defined mathematically in several respects. Nevertheless, it is very important to analyze (6-13) in detail because it is the only known explicit expression for the nontrivial S-matrix in the quantum field theory.

The S-matrix element is dependent on certain factors related purely to the quantum-mechanical states of the particles present in the initial and final states. It is convenient, therefore, to drop those "kinematical" factors from the S-matrix element because they are not essential. Furthermore, when we work in the momentum space, that is, when we take the fourier transform, it is customary to omit a four-dimensional δ-function expressing the overall conservation of the 4-momentum, which is due to the translational invariance of the theory. In this way, we obtain the essential part of the S-matrix element; we call it a *transition amplitude*.

The transition amplitude is most conveniently calculated by means of Feynman integrals. They are easily written down according to the Feynman rules [10, 11, 12] if the corresponding Feynman graphs are given. A Feynman graph is constructed as follows. The number of its external lines is equal to the total number of the particles present in the initial and final states. The number M of vertices corresponds to the order of the term in (6-13). If $\mathscr{L}_{\text{int}}(x)$ is a monomial of degree k in fields (its coefficient is called a *coupling constant*), the F-degree of each vertex is equal to k. The lines incident with a vertex should be labeled according to the names of the fields involved in $\mathscr{L}_{\text{int}}(x)$. If $\mathscr{L}_{\text{int}}(x)$ is a polynomial, each vertex is specified by one of its terms in the above-mentioned way, that is, there are various types of vertices. Each of external and internal lines should be incident with a vertex or a pair of vertices so as to be consistent with the above-mentioned labeling. For an M-th order term of (6-13), it is possible to construct many Feynman graphs. It equals a sum of Feynman integrals (apart from a four-dimensional δ-function) over all possible Feynman graphs which involve no vacuum polarization parts (this restriction comes out from being divided by $\langle 0|\tilde{S}|0\rangle$ in the definition of S).

Let p and m be a 4-momentum and a mass of a particle, respectively. If p satisfies the relation $p^2 = m^2$, then it is said to be *on the mass shell*; otherwise it is said to be *off the mass shell*. In the transition amplitude, all the 4-momenta of external particles (i.e., particles associated with external lines) are on the mass shell. It is possible and convenient, however, to extend them to the ones being off the mass shell straightforwardly. The off-the-mass-shell extension of the transition amplitude is called the *extended transition amplitude*. Correspondingly, Feynman integrals also are extended to the quantities off the mass shell. Hereafter, we always use the terminology of the Feynman integral in this extended sense without writing so explicitly.

Let n be the total number of external particles. The transition amplitude is physically meaningful only for $n \geqq 4$ because at least two particles are present in each of the initial and the final state. The transition amplitudes for $n = 4$ and for $n \geqq 5$ are called the *scattering amplitude* and the *production amplitude*,respectively. On the other hand, the extended transition amplitude is meaningful for $n \geqq 2$. The extended transition amplitudes for $n = 2$ and for $n = 3$ are called the *self-energy* and the *vertex function*, respectively. Those for $n = 4$ and for $n \geqq 5$ are called the *extended scattering amplitude* and the *extended production amplitude*, respectively.

6-2 Definition of the Feynman Integral

We first consider the simplest case in which the interaction Lagrangian density $\mathscr{L}_{\text{int}}(x)$ is a polynomial in spinless-particle fields alone and involves no derivatives of them. It is convenient to work in the momentum space, that is, we take the four-dimensional transform. Then the *Feynman propagator* for a spinless particle labeled by l is given by

$$-i/(m_l^2 - k_l^2 - i\varepsilon_l) \tag{6-15}$$

where m_l and k_l are its mass and 4-momentum, respectively, and ε_l is a positive constant which should tend to zero at the final stage of the calculation.

We consider a connected Feynman graph G, which has N internal lines and M vertices. The *Feynman integral associated with* G is a formal multiple integral

$$F_G(p) \equiv \prod_{l\in|G|}\left(\int d^4k_l\right)\frac{\prod_{b\in v(G)}\delta^4\left(\sum_{l\in|G|}[b:l]\,k_l + p_b\right)}{\prod_{l\in|G|}(m_l^2 - k_l^2 - i\varepsilon_l)} \tag{6-16}$$

that is, it is an integral of a product of the Feynman propagators corresponding to all internal lines of G (apart from a numerical factor) such that the 4-momentum is conserved at every vertex of G. In (6-16), the integration range goes over a $4N$-dimensional real euclidean space, and $\delta^4(k)$ indicates a four-dimensional δ-function, i.e. $\prod_{j=0}^{3}\delta(k^{(j)})$; p_b is an algebraic sum of external momenta outgoing from the vertex b, where an *external momentum* is a 4-momentum of a particle associated with an external line.† We call k_l an *internal momentum* and likewise m_l an *internal mass*.

From (2-5), we have

$$\sum_{b\in v(G)} \Big(\sum_{l\in|G|} [b:l]\, k_l + p_b \Big) = \sum_{b\in v(G)} p_b \tag{6-17}$$

Hence (6-16) is proportional to $\delta^4\Big(\sum_{b\in v(G)} p_b \Big)$, that is, we have the overall conservation of the 4-momentum. The number of independent four-dimensional δ-functions involving integration variables is rank $\mathscr{A} = M - 1$. After carrying out those $M - 1$ trivial integrations, the number of the remaining four-dimensional integrations is $N - (M - 1) = \mu$, the connectivity of G. It is convenient to decompose k_l into an integration momentum $\hat{k}_l$ and a constant one q_l in such a way that

$$k_l = \hat{k}_l + q_l \quad (l \in |G|) \tag{6-18}$$

$$\sum_{l\in|G|} [b:l]\, q_l + p_b = 0 \quad (b \in v(G) - a) \tag{6-19}$$

where a is a particular vertex. Then

$$\prod_{b\in v(G)} \delta^4\Big(\sum_{l\in|G|} [b:l]\, k_l + p_b \Big) = \delta^4\Big(\sum_{b\in v(G)} p_b \Big) \prod_{b\neq a} \delta^4\Big(\sum_{l\in|G|} [b:l]\, \hat{k}_l \Big) \tag{6-20}$$

Because of (3-18) and (3-7), we have

$$\prod_{b\neq a} \delta^4\Big(\sum_{l\in|G|} [b:l]\, \hat{k}_l \Big) = \prod_{j=\mu+1}^{N} \delta^4\Big(\sum_{l\in|G|} [S_j:l]\, \hat{k}_l \Big) \tag{6-21}$$

where $\{S_{\mu+1}, \ldots, S_N\}$ is the f-set of cut-sets corresponding to a tree $T = \{l_{\mu+1}, \ldots, l_N\}$ in the sense of Theorem 2-29. Furthermore, Theorem 3-5 implies

$$\sum_{l\in|G|} [S_j:l]\, \hat{k}_l = \hat{k}_{l_j} - \sum_{i=1}^{\mu} [C_i:l_j]\, \hat{k}_{l_i} \quad (j = \mu + 1, \ldots, N) \tag{6-22}$$

where $\{C_1, \ldots, C_\mu\}$ is the f-set of circuits corresponding to a chord set $T^* = \{l_1, \ldots, l_\mu\}$. By means of (6-18)–(6-22), (6-16) becomes

$$F_G(p) = \delta^4\Big(\sum_{b\in v(G)} p_b \Big) \prod_{l'\in T^*} \Big(\int d^4\hat{k}_{l'} \Big) \prod_{l\in|G|} [m_l^2 - (\hat{k}_l + q_l)^2 - i\varepsilon_l]^{-1} \tag{6-23}$$

where the momenta $\hat{k}_l$ for $l \in T$ are redefined now by

$$\hat{k}_l \equiv \sum_{i=1}^{\mu} [C_i:l]\, \hat{k}_{l_i} \tag{6-24}$$

† We also call p_b an external momentum later.

The μ momenta $\hat{k}_l$ for $l \in T^*$ are called *basic momenta*; (6-24) becomes a trivial identity for any one of basic momenta.

We note that the constant momenta q_l are not uniquely defined by (6-18) and (6-19). Indeed, for any chord set T^*, we can arbitrarily choose q_l for any $l \in T^*$, because then the $M - 1$ simultaneous equations

$$\sum_{l \in T} [b:l]\, q_l = -p_b - \sum_{l \in T^*} [b:l]\, q_l \qquad (b \neq a) \tag{6-25}$$

can be solved with respect to $q_l\,(l \in T)$ owing to det $\mathscr{A}_T^a \neq 0$.

The Feynman integral (6-23) will, in general, not well defined mathematically owing to the following four possible reasons.

(1) Since Feynman propagators should be regarded as distributions (see Appendix A), their product as in (6-23) is not necessarily well defined. This difficulty can be bypassed by regarding ε_l to be finite temporarily.

(2) The integral is at best conditionally convergent because $(\hat{k}_l + q_l)^2$ can remain small for very large $\hat{k}_l^{(\nu)}(\nu = 0, 1, 2, 3)$. Therefore the way of carrying out the integrations has to be specified explicitly. The most satisfactory way is to convert it to a Feynman-parametric integral as done in the next section.

(3) Even if the difficulty (2) is removed, (6-23) may be divergent owing to the contribution from very large $\hat{k}_l^{(\nu)}$. This divergence is known as *ultraviolet divergence*. This difficulty is removed by renormalization as explained in Subsection 10-2. At present, we bypass it by replacing (6-23) by

$$F_G(p, \lambda) \equiv \delta^4\Big(\sum_{b \in v(G)} p_b\Big) \prod_{l' \in T^*} \Big(\int d^4\hat{k}_{l'}\Big) \prod_{l \in |G|} [m_l^2 - (\hat{k}_l + q_l)^2 - i\varepsilon_l]^{-\lambda_l} \tag{6-26}$$

where Re $\lambda_l > 0$ should be chosen in such a way that (6-26) is free from ultraviolet divergence.

(4) When some masses are zero, (6-23) may be divergent owing to the contribution from vanishing $k_l = \hat{k}_l + q_l$. This divergence is known as *infrared divergence*, which cannot be removed in the transition amplitude but automatically cancels out in the transition probability. We avoid this difficulty here by supposing $m_l > 0$ for all $l \in |G|$.

Finally, we briefly mention the case in which $\mathscr{L}_{\text{int}}(x)$ involves nonzero-spin-particle fields and/or derivatives of fields. The Feynman propagator for a Dirac particle is given by (6-15) multiplied by $m_l + \gamma k_l$. [Any line corresponding to a fermion such as a Dirac particle has to be given an intrinsic direction. The Feynman propagator is defined according to this direction.] Likewise, the Feynman propagator for a higher-spin particle can be obtained from (6-15) by multiplying a certain (matrix) polynomial in k_l. If $\mathscr{L}_{int}(x)$ involves nonfield factors such as constant matrices and differentials $(\partial/\partial x)$, then for each vertex b we should introduce a matrix factor or a polynomial in $k_l (l \in S[b])$. We need some additional rules for treating matrices, but we omit them. After all, we should multiply the integrand of (6-16) or (6-23) by a certain matrix polynomial, $P(k)$, in $k_l = \hat{k}_l + q_l \, (l \in |G|)$, which may also depend on external momenta. It is difficult to compute a product of matrices involved in $P(k)$ generally, but recently a systematic method of calculating a product of γ matrices has been proposed [13].

§7 FEYNMAN-PARAMETRIC INTEGRAL

7-1 Derivation of the Feynman-Parametric Integral

It is important from both theoretical and practical points of view to convert the Feynman integral defined in Subsection 6-2 into a Feynman-parametric integral, whose integration region is compact. The Feynman-parametric integral in the general form is derived on the basis of the following *generalized Feynman identity.*

THEOREM 7-1.

$$\frac{1}{\prod_{j=1}^{n} A_j^{\lambda_j}} = \frac{\Gamma(\lambda_0)}{\prod_{j=1}^{n} \Gamma(\lambda_j)} \prod_{j=1}^{n} \left(\int_0^\infty d\alpha_j \right) \times \frac{\delta\left(1 - \sum_{j=1}^{n} \alpha_j\right) \prod_{j=1}^{n} \alpha_j^{\lambda_j - 1}}{\left(\sum_{j=1}^{n} \alpha_j A_j \right)^{\lambda_0}} \tag{7-1}$$

where $A_j \neq 0$ and $\operatorname{Re} \lambda_j > 0$ for all j and $\lambda_0 \equiv \sum_{j=1}^{n} \lambda_j$. Especially,

$$\frac{1}{\prod_{j=1}^{n} A_j} = (n-1)! \prod_{j=1}^{n} \left(\int_0^\infty d\alpha_j \right) \frac{\delta\left(1 - \sum_{j=1}^{n} \alpha_j\right)}{\left(\sum_{j=1}^{n} \alpha_j A_j \right)^n} \tag{7-2}$$

In the above, we may replace the upper limit, ∞, of each integration by 1 because of the δ-function.

Proof We can prove (7-1) by mathematical induction with respect to n by using Euler's formula

$$\int_0^1 d\alpha \frac{\alpha^{a-1}(1-\alpha)^{b-1}}{[\alpha A + (1-\alpha) B]^{a+b}} = \frac{\Gamma(a)\,\Gamma(b)}{\Gamma(a+b)} \cdot \frac{1}{A^a B^b} \tag{7-3}$$

with $\operatorname{Re} a > 0$ and $\operatorname{Re} b > 0$. q.e.d.

Our procedure is first to apply (7-1) to (6-26) or (7-2) to (6-23), second to interchange the order of integrations,† and finally to carry out the momentum integrations explicitly. After the first and second steps mentioned above, (6-26) becomes

$$F_G(p, \lambda) = \delta^4\Big(\sum_b p_b\Big)\Big[\prod_{l\in|G|} \Gamma(\lambda_l)\Big]^{-1} \int_\Delta d^{N-1}\alpha \prod_{l\in|G|} \alpha_l^{\lambda_l-1} H(q, \alpha, \lambda) \tag{7-4}$$

Here

$$\int_\Delta d^{N-1}\alpha \equiv \prod_{l\in|G|}\left(\int_0^\infty d\alpha_l\right)\delta\Big(1 - \sum_{l\in|G|}\alpha_l\Big) \tag{7-5}$$

$$H(q, \alpha, \lambda) \equiv \Gamma(\lambda_0) \prod_{l\in T^*}\left(\int d^4\hat{k}_l\right)[\Phi(\hat{k}, q, \alpha) - i\varepsilon]^{-\lambda_0} \tag{7-6}$$

where

$$\lambda_0 \equiv \sum_{l\in|G|} \lambda_l \tag{7-7}$$

$$\varepsilon \equiv \sum_{l\in|G|} \alpha_l\varepsilon_l \tag{7-8}$$

and

$$\Phi(\hat{k}, q, \alpha) \equiv \sum_{l\in|G|} \alpha_l[m_l^2 - (\hat{k}_l + q_l)^2] = -\sum_l \alpha_l\hat{k}_l^2 - 2\sum_l \alpha_l q_l\hat{k}_l + L(q, \alpha) \tag{7-9}$$

with

$$L(q, \alpha) \equiv \sum_{l\in|G|} \alpha_l(m_l^2 - q_l^2) \tag{7-10}$$

To proceed to the third step, we substitute (6-24) for $\hat{k}_l$ in (7-9):

$$\Phi = -\sum_{i=1}^{\mu}\sum_{j=1}^{\mu}\Big(\sum_{l\in|G|}[C_i : l]\,[C_j : l]\,\alpha_l\Big)\hat{k}_{l_i}\hat{k}_{l_j} - 2\sum_{j=1}^{\mu} \hat{q}_{C_j}\hat{k}_{l_j} + L \tag{7-11}$$

where

$$\hat{q}_C \equiv \sum_{l\in|G|} [C : l]\,\alpha_l q_l \tag{7-12}$$

It is convenient to reexpress (7-11) in the matrix notation. For this purpose, hereafter we introduce the vector notation. We denote the column vectors $(\hat{k}_{l_1}, \ldots, \hat{k}_{l_\mu})^t$ and $(\hat{q}_{C_1}, \ldots, \hat{q}_{C_\mu})^t$ by $\hat{\boldsymbol{k}}_{T^*}$ and $\hat{\boldsymbol{q}}_{T^*}$, respectively. Then (7-11) is rewritten as

$$\Phi = -(\hat{\boldsymbol{k}}_{T^*})^t\,\mathscr{U}_{T^*}\hat{\boldsymbol{k}}_{T^*} - 2(\hat{\boldsymbol{q}}_{T^*})^t\,\hat{\boldsymbol{k}}_{T^*} + L \tag{7-13}$$

where $\mathscr{U}_{T^*}$ is defined by (3-23).

† We do not prove the adequacy of this interchange because it should rather be regarded as the definition of the Feynman integral (see Section 6).

THEOREM **7-2** All the eigenvalues of the symmetric matrix $\mathscr{U}_{T*}$ are positive for $\alpha_l > 0$ ($l \in |G|$).

Proof Any principal minor of $\mathscr{U}_{T*}$ is expressed by a derivative of $U = \det \mathscr{U}_{T*}$ with respect to α_l for $l \in I$ ($I \subset T^*$), which is positive definite because of (3-25). q.e.d. [We can also prove this theorem by noting that $\sum_l \alpha_l \hat{k}_l^2$ is a positive-definite quadratic form if we suppose $\hat{k}_l$ to be real numbers.]

Let η be a very small positive number and

$$\int_{\Delta(\eta)} d^{N-1}\alpha \equiv \prod_{l\in|G|}\left(\int_\eta^\infty d\alpha_l\right)\delta\left(1-\sum_{l\in|G|}\alpha_l\right) \tag{7-14}$$

Then $(\mathscr{U}_{T*})^{-1}$ exists for $\alpha_l \geqq \eta$ ($l \in |G|$). Let

$$\hat{k}_{T*} = k'_{T*} - (\mathscr{U}_{T*})^{-1}\,\hat{q}_{T*} \tag{7-15}$$

then

$$\Phi = -(k'_{T*})^t\,\mathscr{U}_{T*}k'_{T*} + (\hat{q}_{T*})^t\,(\mathscr{U}_{T*})^{-1}\,\hat{q}_{T*} + L \tag{7-16}$$

The Jacobian of the transformation (7-15) is of course unity. Next, we transform $\mathscr{U}_{T*}$ into a diagonal matrix $\mathscr{D}_{T*}$:

$$\begin{aligned} k'_{T*} &= \mathscr{L}_{T*}k''_{T*} \\ (\mathscr{L}_{T*})^{-1}\,\mathscr{U}_{T*}\mathscr{L}_{T*} &= \mathscr{D}_{T*} \end{aligned} \tag{7-17}$$

with $(\mathscr{L}_{T*})^t = (\mathscr{L}_{T*})^{-1}$. Then (7-16) becomes

$$\Phi = -(k''_{T*})^t\,\mathscr{D}_{T*}k''_{T*} + (\hat{q}_{T*})^t\,(\mathscr{U}_{T*})^{-1}\,\hat{q}_{T*} + L \tag{7-18}$$

Here

$$\det \mathscr{D}_{T*} = \det \mathscr{U}_{T*} = U(\alpha) \tag{7-19}$$

because of (3-30).

THEOREM **7-3**

$$\int \frac{d^4k}{(a-bk^2)^\lambda} = \frac{\pi^2 i}{(\lambda-1)(\lambda-2)}\cdot\frac{1}{b^2a^{\lambda-2}} \tag{7-20}$$

where $\operatorname{Im} a < 0$, $b > 0$, and $\operatorname{Re}\lambda > 2$, provided that the $k^{(0)}$ integration is carried out first.

Proof Since the integrand is holomorphic in the first and third quadrants in the $k^{(0)}$ plane, we can rotate the contour into the imaginary axis by means of Cauchy's theorem. We then have an integral involving a four-dimensional euclidean vector, which can be easily calculated by considering it in polar coordinates. q.e.d.

By means of (7-20), we can straightforwardly carry out the integrations over $k_l''(l \in T^*)$; (7-6) becomes

$$H(q, \alpha, \lambda) = \Gamma(\lambda_0 - 2\mu)\, (\pi^2 i)^\mu / U^2(\alpha)\, [V(q, \alpha) - i\varepsilon]^{\lambda_0 - 2\mu} \tag{7-21}$$

where from (7-18) with (7-10) we have

$$V(q, \alpha) \equiv \sum_{l \in |G|} \alpha_l (m_l^2 - q_l^2) + (\hat{q}_{T^*})^t (\mathscr{U}_{T^*})^{-1} \hat{q}_{T^*} \tag{7-22}$$

The above result can be summarized as follows.

THEOREM **7-4**

$$F_G(p, \lambda) = \delta^4\Big(\sum_b p_b\Big) (\pi^2 i)^\mu f_G(p, \lambda) \tag{7-23}$$

with

$$f_G(p, \lambda) \equiv \frac{\Gamma(\lambda_0 - 2\mu)}{\prod_{l \in |G|} \Gamma(\lambda_l)} \lim_{\eta \to +0} \int_{\Delta(\eta)} d^{N-1}\alpha \frac{\sum_{l \in |G|} \alpha_l^{\lambda_l - 1}}{U^2 (V - i\varepsilon)^{\lambda_0 - 2\mu}} \tag{7-24}$$

If we take Re $\lambda_l > 2$, then as seen in Subsection 10-1, the $\eta \to +0$ limit exists, that is, we may replace $\lim_{\eta \to +0} \int_{\Delta(\eta)}$ by $\int_{\Delta}$. The genuine Feynman integral $F_G(p)$ is formally $F_G(p, 1)$ for $N > 2\mu$, where $\lambda = 1$ means $\lambda_l = 1$ for all $l \in |G|$. We write $f_G(p) \equiv f_G(p, 1)$.

In deriving (7-24) we have employed a particular chord set T^*. Since the original Feynman integral (6-16) is, however, independent of T^*, (7-24), and therefore V, should be independent of it. We can check this property directly.

Let $\hat{k}_J$ be a column vector formed by μ independent integration momenta, so that $\hat{k}_{T^*} = \mathscr{J} \hat{k}_J$ with det $\mathscr{J} \neq 0$. Then $\mathscr{U}_{T^*}$ and $\hat{q}_{T^*}$ in (7-16) are replaced by $\mathscr{U}_J \equiv \mathscr{J}^t \mathscr{U}_{T^*} \mathscr{J}$ and $\hat{q}_J \equiv \mathscr{J}^t \hat{q}_{T^*}$, respectively; the $\hat{k}_J$ integrations may be done by the same manipulations as used above. We observe that the extra factor $(\det \mathscr{U}_J)^2/(\det \mathscr{U}_{T^*})^2$ in the denominator exactly cancels the Jacobian $(\det \mathscr{J})^4$ and that

$$(\hat{q}_J)^t (\mathscr{U}_J)^{-1} \hat{q}_J = (\hat{q}_{T^*})^t (\mathscr{U}_{T^*})^{-1} \hat{q}_{T^*} \tag{7-25}$$

7-2 Topological Formulas for the *V* Function

The central importance in this chapter is to derive topological formulas for V defined by (7-22).

THEOREM **7-5** For any values of q_l ($l \in |G|$) which may not satisfy (6-19), V is expressed as follows:

(a) *Circuit representation* [3]

$$V = \sum_{l\in|G|} \alpha_l(m_l^2 - q_l^2) + U^{-1} \sum_{C\in\mathbf{C}} U_C\Big(\sum_{l\in|G|} [C:l]\,\alpha_l q_l\Big)^2 \tag{7-26}$$

(b) *Cut-set representation* [7, 8]

$$V = \sum_{l\in|G|} \alpha_l m_l^2 - U^{-1} \sum_{S\in\mathbf{S}} W_S\Big(\sum_{l\in|G|} [S:l]\, q_l\Big)^2 \tag{7-27}$$

Proof of (7-26) [3] Comparing (7-22) with (7-26) together with (7-12), we see that we have only to prove

$$\sum_{i=1}^{\mu} \sum_{j=1}^{\mu} [\mathscr{U}_{T*}]^{i,j}\, \hat{q}_{C_i}\hat{q}_{C_j} = \sum_{C\in\mathbf{C}} U_C\hat{q}_C^2 \tag{7-28}$$

with $[\mathscr{U}_{T*}]^{i,j} \equiv [\mathscr{U}_{T*}]^{C_i,C_j}$. In the following we prove the identity

$$\sum_{i=1}^{\mu} \sum_{j=1}^{\mu} [\mathscr{U}_{T*}]^{i,j}\Big(\sum_{l} [C_i:l]\,\bar{q}_l\Big)\Big(\sum_{l'} [C_j:l']\,\bar{q}_{l'}\Big) = \sum_{C\in\mathbf{C}} U_C\Big(\sum_{l} [C:l]\,\bar{q}_l\Big)^2 \tag{7-29}$$

for arbitrary $\bar{q}_l$ ($l \in |G|$); (7-28) follows from (7-29) by setting $\bar{q}_l = \alpha_l q_l$.

Now, we take an arbitrary circuit C and put $\alpha(C) = 0$ (i.e., $\alpha_l = 0$ for all $l \in C$). Since the left-hand side of (7-29) is independent of T^* as shown in Subsection 7-1, we may choose an f-set of circuits $\{C_1, \ldots, C_\mu\}$ which contains C, say $C_1 = C$. Then Theorem 3-13 shows

$$[\mathscr{U}_{T*}]^{1,1}_{\alpha(C)=0} = U_C \tag{7-30}$$

while

$$[\mathscr{U}_T]^{i,j}_{\alpha(C)=0} = 0 \quad \text{unless} \quad i = j = 1 \tag{7-31}$$

because all elements of its first row or column vanish. Therefore the left-hand side of (7-29) reduces to $U_C\Big(\sum_l [C:l]\bar{q}_l\Big)^2$ for $\alpha(C) = 0$.

Next, for any $C' \neq C$ we prove

$$U_{C'} = 0 \quad \text{for} \quad \alpha(C) = 0 \tag{7-32}$$

Since $U_{C'}$ is a chord-set product sum in G/C', any term of it corresponds to a chord set T'^* in G/C'. Since $C \not\subset C'$, $C - C'$ contains a circuit in G/C' because of Theorem 2-2 (c). Hence $T'^* \cap C \neq \emptyset$, from which (7-32) follows.

Let R be the difference between the left-hand and the right-hand side of (7-29). From the above arguments, R vanishes when $\alpha(C) = 0$ for any circuit C. However, since R is a polynomial in α_l ($l \in |G|$) of degree $\mu - 1$, if $R \not\equiv 0$ then R contains a term of the form $k \prod_{j=1}^{\mu-1} \alpha_{l_j}$ ($k \neq 0$) for some lines $l_1, \ldots, l_{\mu-1}$, which may not be all distinct. Let $I \equiv \bigcup_{j=1}^{\mu-1} \{l_j\}$; then $N(I^*) \geqq M$, where $I^* \equiv |G| - I$. Hence I^* includes a circuit C because of Theorems 2-6 (b) and 2-5 (d), so that $I \cap C = \emptyset$. The existence of such a term is inconsistent with the property of R derived above. Therefore $R \equiv 0$. q.e.d.

Before proving (7-27), we prove the following identities.

THEOREM **7-6**

(a)
$$\partial U / \partial \alpha_l = \sum_{C \ni l} U_C \tag{7-33}$$

(b)
$$\alpha_l \cdot U|_{\alpha_l = 0} = \sum_{S \ni l} W_S \tag{7-34}$$

(c)
$$\sum_{C \in \boldsymbol{C}} [C:l]\,[C:l']\,\alpha_l \alpha_{l'} U_C = -\sum_{S \in \boldsymbol{S}} [S:l]\,[S:l']\,W_S \quad \text{for} \quad l \neq l' \tag{7-35}$$

Proof (a). It is convenient to translate (7-33) into the inverse-Feynman-parametric form:

$$\sum_{T \not\ni l} \prod_{k \in T + l} \beta_k = \sum_{C \ni l} \sum_{T^0 \supset C} \prod_{k \in T^0} \beta_k \tag{7-36}$$

where T is a tree, T^0 being a pseudotree. Given a tree $T \not\ni l$, $T^0 \equiv T + l$ is a pseudotree such that $l \in C \subset T^0$ where C is a circuit. Conversely, given a pseudotree $T^0 \supset C \ni l$, $T \equiv T^0 - l$ is a tree which does not contain l (see Theorem 2-6 (b)). Thus (7-36) holds identically.

(b). We rewrite (7-34) as

$$\sum_{T \ni l} \prod_{k \in T - l} \beta_k = \sum_{S \ni l} \sum_{T^2 \in \boldsymbol{T}^2(S)} \prod_{k \in T^2} \beta_k \tag{7-37}$$

where T and S denote a tree and a cut-set, respectively. Given a tree $T \ni l$, there exists a unique cut-set S such that $T \cap S = \{l\}$ because of Theorem 2-29. Then the 2-tree $T^2 \equiv T - l$ is disjoint from S. Conversely, given a 2-tree $T^2 \in \boldsymbol{T}^2(S)$ with $S \ni l$, $T \equiv T^2 + l$ is a tree because of Theorem 2-6 (b).

(c). The equality which we have to prove is

$$\sum_{C} [C:l]\,[C:l'] \sum_{T^0 \supset C} \prod_{k \in T^0 - l - l'} \beta_k = -\sum_{S} [S:l]\,[S:l'] \sum_{T^2 \in \boldsymbol{T}^2(S)} \prod_{k \in T^2} \beta_k$$
$$\text{for} \quad l \neq l' \tag{7-38}$$

Given a 2-tree $T^2 \in \boldsymbol{T}^2(S)$ with $S \ni l, l'$, $T^0 \equiv T^2 + l + l'$ is a pseudotree because $T^2 + l$ is a tree. Therefore, T^0 includes a unique circuit C, and $l, l' \in C$ because both $T^0 - l$ and $T^0 - l'$ are trees. Hence $S \cap T^2 = \emptyset$ implies $C \cap S = \{l, l'\}$, whence (2-12) reduces to

$$[C:l]\,[S:l] + [C:l']\,[S:l'] = 0 \tag{7-39}$$

that is,

$$[C:l]\,[C:l'] = -[S:l]\,[S:l'] \tag{7-40}$$

Conversely, given a pseudotree $T^0 \supset C \ni l, l'$, since $T^0 - l$ is a tree, from Theorem 2-29 there exists a unique cut-set S such that $(T^0 - l) \cap S = \{l'\}$. Hence $T^2 = T^0 - l - l'$ is a 2-tree, and $l, l' \in S$ because $N(S \cap C)$ is even. Thus T^0 and $T^2 \in \boldsymbol{T}^2(S)$ are in one-to-one correspondence. q.e.d.

Proof of (7-27) Since

$$U = \frac{\partial U}{\partial \alpha_l}\alpha_l + U|_{\alpha_l=0} \quad \text{for any } l \tag{7-41}$$

we have

$$U \sum_{l\in|G|} \alpha_l q_l^2 = \sum_{l\in|G|} \Big(\sum_{C\ni l} U_C\Big) \alpha_l^2 q_l^2 + \sum_{l\in|G|} \alpha_l \,(U|_{\alpha_l=0})\, q_l^2 \tag{7-42}$$

with the aid of (7-33). Hence the coefficient of q_l^2 in (7-26) is

$$-U^{-1} \sum_{l\in|G|} \alpha_l\, U|_{\alpha_l=0} \tag{7-43}$$

and that of $q_l q_{l'}$ $(l \neq l')$ is

$$U^{-1} \sum_C [C:l]\,[C:l']\,\alpha_l \alpha_{l'} U_C \tag{7-44}$$

On account of (7-34) and (7-35), therefore, we immediately obtain (7-27). q.e.d.

THEOREM **7-7** In terms of external momenta V is expressed as follows:

(a) *Vertex representation* [3, 14, 15]

$$V = \sum_{l\in|G|} \alpha_l m_l^2 + U^{-1} \sum_{\{b,c\}\subset g} W^{(b|c)} p_b p_c \tag{7-45}$$

(b) *Modified cut-set representation* [4, 5, 6]

$$V = \sum_{l\in|G|} \alpha_l m_l^2 - U^{-1} \sum_{(h|g-h)} W^{(h|g-h)} \Big(\sum_{b\in h} p_b\Big)^2 \tag{7-46}$$

where g is the set of all external vertices (i.e., $p_b = 0$ for any $b \notin g$) and the summation $\sum_{(h|g-h)}$ runs over all nontrivial divisions of g.

Proof From (7-27), (7-46) immediately follows by using the momentum conservation

$$\sum_{l\in|G|} [S:l]\, q_l = \pm \sum_{b\in h} p_b \quad \text{for} \quad S \in \mathcal{S}(h \mid g - h) \tag{7-47}$$

and (3-43) (without tildas). Here (7-47) is derived by means of (2-10) and (6-19).

The equivalence between (7-45) and (7-46) are shown as follows:

$$\begin{aligned}\sum_{\{b,c\}\subset g} W^{(b|c)}\, p_b p_c &= \sum_{\{b,c\}\subset g} \sum_{\substack{h\ni b \\ g-h\ni c}} W^{(h|g-h)}\, p_b p_c \\ &= \sum_{(h|g-h)} W^{(h|g-h)} \Big(\sum_{b\in h} p_b\Big)\Big(\sum_{c\in g-h} p_c\Big) \\ &= -\sum_{(h|g-h)} W^{(h|g-h)} \Big(\sum_{b\in h} p_b\Big)^2 \end{aligned}\tag{7-48}$$

with the aid of (3-46) (without tildas) in the first step. q.e.d.

We can also present a direct proof of (7-45) in the following way.

We introduce the following matrix $\mathscr{V}_{T*}$ of order $\mu + 1$:

$$\mathscr{V}_{T*} \equiv \begin{bmatrix} \sum_l \alpha_l q_l^2 & (q_{T*})^t \\ \hat{q}_{T*} & \mathscr{U}_{T*} \end{bmatrix} \tag{7-49}$$

Then from the definition (7-22) of V we have

$$U\Big(\sum_l \alpha_l m_l^2 - V\Big) = U \sum_l \alpha_l q_l^2 - \sum_{l,l'\in T*} [\mathscr{U}_{T*}]^{l,l'} \hat{q}_l \hat{q}_{l'} = \det \mathscr{V}_{T*} \tag{7-50}$$

As remarked in Subsection 6-2, we may put $q_l = 0$ for all $l \in T^*$, and then the remainders $q_l\,(l \in T)$ are fixed as linear combinations of external momenta p_b. Furthermore, by means of the overall momentum conservation we have

$$p_b^2 = -p_b \sum_{c\neq b} p_c \tag{7-51}$$

Therefore the quantity in (7-50) can be expanded into

$$\sum_{\{b,c\}\subset v(G)} A_{bc} p_b p_c = \sum_{\{b,c\}\subset g} A_{bc} p_b p_c \tag{7-52}$$

where the coefficients A_{bc} are functions of $\alpha_l\,(l \in |G|)$. Comparing (7-52) and (7-45), we have only to show

$$-A_{bc} = W^{(b|c)} \tag{7-53}$$

Since we can choose external momenta arbitrarily as long as they satisfy the conservation law, we put

$$p_b = -p_c, \ p_a = 0 \quad \text{unless} \quad a = b \quad \text{or} \quad c \tag{7-54}$$

then (7-52) reduces to $-A_{bc}p_b^2$. Accordingly, our task is to show that det $\mathscr{V}_{T^*}$ with (7-54) is equal to $W^{(b|c)}p_b^2$.

Now, the tree T includes a unique path $P \in \boldsymbol{P}(bc)$ because of Theorem 2-5 (c). We consider a Feynman graph $G^{(bc)'}$ which is obtained from G by linking b and c by a new line l_0. Let $\{C_1, \ldots, C_\mu\}$ be the f-set of circuits in G corresponding to the chord set $T^* = \{l_1, \ldots, l_\mu\}$; then since $T^* \cap P = \emptyset$, $\{C_0, C_1, \ldots, C_\mu\}$ is an f-set of circuits in $G^{(bc)'}$, where $C_0 \equiv P + l_0$. We define the orientation of C_0 by setting $[C_0 : l] = [b : l_0]$, and put

$$q_l = [C_0 : l]\, p_b \quad \text{for all} \quad l \in |G| \tag{7-55}$$

Then (7-55) satisfies the conservation law at each vertex and $q_l = 0$ for all $l \in T^*$ as required. On substituting (7-55) in (7-12), we find that (7-49) becomes

$$\det \mathscr{V}_{T^*} = p_b^2 \det \Big(\sum_{l \in |G|} [C_j : l]\,[C_{j'} : l]\, \alpha_l \mid j, j' = 0, 1, \ldots, \mu \Big) \tag{7-56}$$

The matrix in the right-hand side of (7-56) is exactly the $\mathscr{U}$ matrix in $G^{(bc)'}$ with $\alpha_{l_0} = 0$, whence its determinant equals the chord-set product sum in $G^{(bc)}$ because of Theorems 3-10 and 3-19. Therefore

$$\det \mathscr{V}_{T^*} = W^{(b|c)}\, p_b^2 \tag{7-57}$$

because $W^{(b|c)}$ is the U function in $G^{(bc)}$. q.e.d.

7-3 Remarks

In this subsection, we make some remarks on applications of the Feynman-parametric formula. We consider it at the physical value of λ, that is, we consider $F_G(p) = F_G(p, 1)$.

THEOREM **7-8** For every Feynman integral $F_G(p)$, there always exists an interaction Lagrangian density $\mathscr{L}_{\text{int}}(x)$ such that the nonvanishing lowest-order contribution to a certain extended transition amplitude coincides with $F_G(p)$ apart from a constant coefficient.

Proof We define $\mathscr{L}_{\text{int}}(x)$ by

$$\mathscr{L}_{\text{int}}(x) = \sum_{b \in v(G)} g_b \psi_b(x) \prod_{l \in S[b]} \hat{\varphi}_l(x) \tag{7-58}$$

Here $\psi_b(x)$ $(b \in g)$ and $\hat{\varphi}_l(x)$ $(l \in |G|)$ are field operators corresponding to distinct spinless particles, and $\psi_b(x) \equiv 1$ for $b \notin g$; the mass of $\hat{\varphi}_l(x)$ is equal to m_l, and g_b $(b \in v(G))$ denotes a coupling constant. It is easy to show by means of the standard analysis [11] that the nonvanishing lowest-order extended transition amplitude for the process specified by g is exactly given by $F_G(p)$ multiplied by a certain constant. q.e.d.

The proportionality constant is given by

$$i^M\Big(\prod_{b\in v(G)} g_b\Big)\,(2\pi)^{4M}\,[-i/(2\pi)^4]^N \tag{7-59}$$

Here the first factor comes out from the coefficient in (6-13) ($1/M!$ is canceled by the $M!$ permutations† of $x_1, \ldots, x_M$), the second factor is evident from (7-58), the third one is due to the $4M$-dimensional fourier transform, and the last one comes out from (6-15) together with the inverse fourier transform. Hence, with the aid of (7-23) with $\lambda = 1$, the contribution, $T_G(p)$, from G to the extended transition amplitude is given by

$$T_G(p) = i(2\pi)^4\,(4\pi)^{-2\mu}\Big(\prod_b g_b\Big)\,f_G(p) \tag{7-60}$$

As remarked in Subsection 6-1, the S matrix is unitary for the physical values of momenta, that is,

$$SS^\dagger = S^\dagger S = 1 \tag{7-61}$$

On setting $S = 1 + iR$, (7-61) becomes

$$RR^\dagger = R^\dagger R = i(R^\dagger - R) \tag{7-62}$$

or, for two (orthogonal) states $|\alpha\rangle$ and $|\beta\rangle$,

$$\sum_\gamma \langle\beta|R^\dagger|\gamma\rangle\,\langle\gamma|R\alpha\rangle = i\langle\beta|(R^\dagger - R)|\,\alpha\rangle \tag{7-63}$$

One half of the right-hand side of (7-63) is called the *absorptive part* of $\langle\beta\,|\,R\,|\,\alpha\rangle$. If we consider the nonvanishing lowest-order term of $\langle\beta|R|\alpha\rangle$ as in Theorem 7-8, it is given by only one Feynman integral. Then its absorptive part is proportional to $\operatorname{Im} f_G(p)$ $(\varepsilon \to +0)$. The left-hand side in the lowest order is a sum of products of simpler Feynman integrals. Each term of those products is expressed as follows. Let h and $g - h$ be the sets of the external lines of the particles present in the states $|\alpha\rangle$ and $|\beta\rangle$,

† If G is symmetric under some permutations of internal vertices, a factor $1/r$ remains, where r denotes the number of the permutations leaving G invariant.

respectively. For each $S \in \boldsymbol{S}(h \mid g - h)$, we consider an integral $F_G^{(S)}(p)$, which is obtained from (6-23) with $\varepsilon \to +0$ by replacing the propagator (6-15) by†

$$2\pi\, \theta(k_l^{(0)})\, \delta(m_l^2 - k_l^2) \tag{7-64}$$

for all $l \in S$. Then the unitarity relation (7-63) implies

$$2\, \mathrm{Im}\, f_G(p) = \sum_{S \in \boldsymbol{S}(h|g-h)} f_G^{(S)}(p) \tag{7-65}$$

with $F_G^{(S)}(p) \equiv \delta^4\left(\sum_b p_b\right)(\pi^2 i)^\mu f_G^{(S)}(p)$.

From (7-24) together with (A-17), $\mathrm{Im}\, f_G(p)\ (\varepsilon \to +0)$ is formally expressed as follows [3, 14]:

$$\mathrm{Im}\, f_G(p) = \pi \int_\Delta d^{N-1}\alpha\, U^{-2}\, \delta^{(N-2\mu-1)}(-V) \tag{7-66}$$

where for $N \leqq 2\mu$, $\delta^{(N-2\mu-1)}(-V)$ should be replaced by

$$\frac{(-V)^{2\mu-N}}{(2\mu - N)!}\, \theta(-V) \tag{7-67}$$

For $N \leqq 2\mu$, $f_G(p)$ is ultraviolet divergent, but $\mathrm{Im}\, f_G(p)\ (\varepsilon \to +0)$ may sometimes become finite because the θ-function restricts the range of the integrations over α.

Example For the $N = 2$ self-energy graph,

$$\mathrm{Im}\, f_G(p) = \pi \int_0^1 d\alpha\, \theta(\alpha(1-\alpha)\, p^2 - \alpha m_1^2 - (1 - \alpha)\, m_2^2)$$

where two lines are numbered as 1 and 2.

Finally, we make two remarks on the expression for the Feynman-parametric integral; λ is not put equal to 1.

REMARK **7-9** We can eliminate $\delta\left(1 - \sum_l \alpha_l\right)$ from (7-24) without violating the homogeneity in α but at the sacrifice of compactness of the integration region. We multiply $f_G(p, \lambda)$ by

$$1 = \int_0^\infty e^{-\xi}\, d\xi \tag{7-68}$$

† $\theta(z)$ denotes Heaviside's step function $\frac{1}{2}(1 + z/|z|)$.

and make a scale transformation $\alpha_l = \alpha'_l/\xi \ (l \in |G|)$. After carrying out the integration over ξ and dropping primes, one obtains

$$f_G(p, \lambda) = \frac{\Gamma(\lambda_0 - 2\mu)}{\prod_l \Gamma(\lambda_l)} \prod_l \left(\int_0^\infty d\alpha_l \right) \frac{\prod_l \alpha_l^{\lambda_l - 1} \left(\sum_l \alpha_l\right) \exp\left(-\sum_l \alpha_l\right)}{U^2 \, (V - i\varepsilon)^{\lambda_0 - 2\mu}} \tag{7-69}$$

Another way of eliminating $\delta\left(1 - \sum_l \alpha_l\right)$ is as follows. We use the identity

$$(V - i\varepsilon)^{-\varrho} = [i^\varrho/\Gamma(\varrho)] \int_0^\infty d\gamma \, \gamma^{\varrho - 1} \, e^{-i\gamma(V - i\varepsilon)} \tag{7-70}$$

with $\varrho = \lambda_0 - 2\mu$, and make a transformation $\alpha_l = \alpha'_l/\gamma \ (l \in |G|)$. After carrying the integration over γ and dropping primes, one obtains

$$f_G(p, \lambda) = \frac{i^{\lambda_0 - 2\mu}}{\prod_l \Gamma(\lambda_l)} \prod_l \left(\int_0^\infty d\alpha_l \right) \frac{\prod_l \alpha_l^{\lambda_l - 1}}{U^2} \, e^{-iV - \varepsilon} \tag{7-71}$$

Contrary to (7-24), (7-71) with $\lambda = 1$ formally makes sense even for $N \leqq 2\mu$.

Remark **7-10** If momenta are of ν dimensions instead of four, the formula (7-23) with (7-24) is extended to

$$F_G(p, \lambda) = \delta^\nu \left(\sum_b p_b\right) i^\mu \pi^{\frac{1}{2}\nu\mu} \frac{\Gamma(\lambda_0 - \frac{1}{2}\nu\mu)}{\prod_l \Gamma(\lambda_l)} \int_\Delta d^{N-1}\alpha \frac{\prod_l \alpha_l^{\lambda_l - 1}}{U^{\nu/2} (V - i\varepsilon)^{\lambda_0 - \frac{1}{2}\nu\mu}} \tag{7-72}$$

8-1 Inverse-Feynman-parametric integral

In this subsection, we derive the inverse-Feynman-parametric integral [4] in order to demonstrate explicitly that the duality between the circuit and the cut-set representation of V is closely related to that between the momentum space and the position space. For simplicity, we consider the case $\lambda = 1$.

From (6-15), the Feynman propagator in the position space is given by

$$\Delta_F(y, m^2) \equiv \frac{-i}{(2\pi)^4} \int d^4k \frac{e^{-iky}}{m^2 - k^2 - i\varepsilon} \tag{8-1}$$

By inserting

$$\frac{1}{m^2 - k^2 - i\varepsilon} = i \int_0^\infty d\alpha \exp\left[-i\alpha(m^2 - k^2 - i\varepsilon)\right] \tag{8-2}$$

we can carry out the momentum integration in (8-1) by means of the Fresnel formula

$$\int d^4k \exp(i\alpha k^2) = -i(\operatorname{sgn}\alpha)\,\pi^2/\alpha^2 \quad \text{for} \quad \alpha \neq 0, \text{ real} \tag{8-3}$$

By setting $\beta = 1/\alpha$, we obtain

$$\Delta_F(y, m^2) = \frac{-i}{(4\pi)^2} \int_0^\infty d\beta \exp\left[-i\left(\frac{1}{4}\beta y^2 + \frac{m^2 - i\varepsilon}{\beta}\right)\right] \tag{8-4}$$

As before, we consider a connected Feynman graph G, in which all particles are spinless and no derivative couplings are involved. Let

$$\tilde{F}_G(x) \equiv i^N \prod_{l \in |G|} \Delta_F\Big(-\sum_{b \in v(G)} [b:l]\, x_b,\, m_l^2\Big) \tag{8-5}$$

Then from (8-4)

$$\tilde{F}_G(x) = (4\pi)^{-2N} \prod_{l \in |G|} \left(\int_0^\infty d\beta_l\right)$$

$$\times \exp(-i) \sum_{l \in |G|} \left\{\tfrac{1}{4}\beta_l \Big(\sum_{b \in v(G)} [b:l]\, x_b\Big)^2 + (m_l^2 - i\varepsilon_l)/\beta_l\right\} \tag{8-6}$$

By using (8-5) and (8-1) we can easily verify the relation

$$F_G(p) = (2\pi)^{4(\mu-1)} \prod_{b\in v(G)} \left(\int d^4x_b\right) e^{i\sum_b p_b x_b} \tilde{F}_G(x) \tag{8-7}$$

we rewrite the first term in the curly bracket of (8-6) as

$$\begin{aligned} \Omega &\equiv \sum_{l\in|G|} \beta_l \Big(\sum_{b\in v(G)} [b:l]\, x_b \Big)^2 \\ &= \sum_{l\in|G|} \beta_l \Big\{ \sum_{b\neq a} [b:l]\,(x_b - x_a) \Big\}^2 \\ &= \sum_{b,c\neq a} w_{bc} x_b^a x_c^a \end{aligned} \tag{8-8}$$

where $x_b^a \equiv x_b - x_a$ and w_{bc} denotes the (b, c) element of the matrix $\mathscr{W}^a$ defined by (3-22). We denote column vectors $(x_b^a \mid b \in v(G) - a)$ and $(p_b \mid b \in v(G) - a)$ by $\boldsymbol{x}^a$ and $\boldsymbol{p}^a$, respectively. Then (8-8) is rewritten as

$$\Omega = (\boldsymbol{x}^a)^t \mathscr{W}^a \boldsymbol{x}^a \tag{8-9}$$

Now, we calculate (8-7) with (8-6). By interchanging the order of the integrations, we have

$$F_G(p) = \frac{(2\pi)^{4(\mu-1)}}{(4\pi)^{2N}} \prod_{l\in|G|} \left(\int_0^\infty d\beta_l \right) \exp\left(-i \sum_{l\in|G|} \frac{m_l^2 - i\varepsilon_l}{\beta_l} \right) H(p, \beta) \tag{8-10}$$

with

$$\begin{aligned} H(p, \beta) &\equiv \prod_{b\in v(G)} \left(\int d^4x_b\right) \exp i\Big(\sum_{b\in v(G)} p_b x_b - \tfrac{1}{4}\Omega \Big) \\ &= (2\pi)^4\, \delta^4\Big(\sum_b p_b\Big) \prod_{b\neq a} \left(\int d^4x_b^a\right) \\ &\quad \times \exp(-i)\, [\tfrac{1}{4}(\boldsymbol{x}^a)^t \mathscr{W}^a \boldsymbol{x}^a - (\boldsymbol{p}^a)^t\, \boldsymbol{x}^a] \end{aligned} \tag{8-11}$$

As before, we first eliminate the linear term in the square bracket of (8-11), and diagonalize the quadratic form; finally we make use of (8-3) to obtain

$$H(p, \beta) = (2\pi)^4\, \delta^4\Big(\sum_b p_b\Big)\, [(4\pi)^2\, i]^{M-1} \frac{\exp\,[i(\boldsymbol{p}^a)^t\, (\mathscr{W}^a)^{-1}\, \boldsymbol{p}^a]}{(\det \mathscr{W}^a)^2} \tag{8-12}$$

On substituting (8-12) in (8-10), we obtain

$$f_G(p) = i^{N-2\mu} \prod_l \left(\int_0^\infty d\beta_l \right) \frac{1}{\tilde{U}^2}\, e^{-i\tilde{V}_{-\varepsilon}} \tag{8-13}$$

where

$$\tilde{U} = \det \mathscr{W}^a = \sum_{T\in\mathcal{T}} \prod_{l\in T} \beta_l \tag{8-14}$$

$$\tilde{V} \equiv \sum_{l\in|G|} m_l^2 \beta_l^{-1} - (\boldsymbol{p}^a)^t (\mathscr{W}^a)^{-1} \boldsymbol{p}^a \tag{8-15}$$

$$\varepsilon \equiv \sum_l \varepsilon_l \beta_l^{-1} \tag{8-16}$$

On account of Theorems 3-16 and 3-18, (8-15) can be rewritten as

$$\tilde{V} = \sum_l m_l^2 \beta_l^{-1} - \tilde{U}^{-1} \Big[\sum_{b\in v(G)-a} \tilde{W}^{(a|b)} p_b^2 + 2 \sum_{\{b,c\}\subset v(G)-a} \tilde{W}^{(a|bc)} p_b p_c \Big] \tag{8-17}$$

Furthermore, by means of (3-46) it is straigthforward to rewrite (8-17) as

$$\tilde{V} = \sum_l m_l^2 \beta_l^{-1} - \tilde{U}^{-1} \sum_{u\subset v(G)-a} \tilde{W}^{(u|u^*)} \Big(\sum_{b\in u} p_b \Big)^2 \tag{8-18}$$

Since the summation over sets of vertices which do not contain a particular vertex is equivalent to that over all possible divisions $(u \mid u^*)$ of $v(G)$, (8-18) is nothing but the modified cut-set representation in the inverse-Feynman-parametric form. [We may rewrite (8-18) as in (7-46) by using $p_b = 0$ for $b \notin g$.] It is evident that (8-13) reduces to (7-71) with $\lambda = 1$ by setting $\beta_l = 1/\alpha_l \, (l \in |G|)$.

The above method of deriving (8-13) is not completely dual to the one presented in Section 7 because x_b is associated with a vertex, which has no dual concept in general. The dual of the internal momentum k_l is $y_l = -\sum_b [b:l]\, x_b = -\sum_{b\neq a} [b:l]\, x_b^a$. Let T be a tree and $\boldsymbol{y}_T$ be a column vector $(y_l \mid l \in T)$; then

$$\boldsymbol{y}_T = -(\mathscr{A}_T^a)^t \, \boldsymbol{x}^a \tag{8-19}$$

By transforming $\boldsymbol{x}^a$ into $\boldsymbol{y}_T$, (8-11) becomes

$$H(p, \beta) = (2\pi)^4 \, \delta^4\Big(\sum_b p_b\Big) \prod_{l\in T} \Big(\int d^4 y_l\Big)$$
$$\times \exp(-i) \, [\tfrac{1}{4}(\boldsymbol{y}_T)^t \, \tilde{\mathscr{U}}_T \boldsymbol{y}_T - (\tilde{\boldsymbol{q}}_T)^t \, \boldsymbol{y}_T] \tag{8-20}$$

with

$$\tilde{\boldsymbol{q}}_T \equiv \mathscr{S}_T \boldsymbol{q}, \quad \boldsymbol{q} = (q_l \mid l \in |G|) \tag{8-21}$$

where we have made use of (3-7), (3-18), (3-22), (3-24), and (6-19). From (8-20) we again obtain (8-13) with

$$\tilde{U} = \det \tilde{\mathscr{U}}_T = \sum_{T\in\mathcal{T}} \prod_{l\in T} \beta_l \tag{8-22}$$

$$\tilde{V} = \sum_l m_l^2 \beta_l^{-1} - (\tilde{\boldsymbol{q}}_T)^t \, \tilde{\mathscr{U}}_T^{-1} \, \tilde{\boldsymbol{q}}_T \tag{8-23}$$

We can prove the cut-set representation of $\tilde{V}$ from (8-23) by using (3-47) in quite an analogous way to the derivation of the circuit representation of V presented in Subsection 7-2. Thus the duality between the two formulas is almost complete.

8-2 Position-Space Feynman-Parametric Integral

The position-space Feynman propagator (8-1) is a somewhat complicated function of y, and therefore it is not convenient to work in the position space directly. From its analyticity in y^2, however, one can show that $\Delta_F(y, m^2)$ $(\varepsilon \to +0)$ has the following spectral representation:

$$\Delta_F(y, m^2) = \frac{1}{\pi} \int_{-0}^{\infty} d\varrho^2 \frac{\bar{\Delta}(\varrho^2, m^2)}{\varrho^2 - y^2 + i\hat{\varepsilon}} \qquad (\hat{\varepsilon} \to +0) \qquad (8\text{-}24)$$

where $\bar{\Delta}(y^2, m^2)$ is the four-dimensional inverse fourier transform of the principal value of $1/(m^2 - k^2)$, that is,

$$\bar{\Delta}(\varrho^2, m^2) = \frac{\delta(\varrho^2)}{4\pi} - \frac{\theta(\varrho^2)}{8\pi} \frac{m}{\varrho} J_1(\varrho m) \qquad (8\text{-}25)$$

J_1 being a Bessel function. The formula of the Hankel transform yields the orthogonality relation of $\bar{\Delta}$:

$$\int_{-0}^{\infty} dm^2\, \bar{\Delta}(\varrho^2, m^2)\, \bar{\Delta}(m^2, \varkappa^2) = (1/16\pi^2)\, \delta(\varrho^2 - \varkappa^2) \qquad (\varrho^2, \varkappa^2 \geqq 0) \quad (8\text{-}26)$$

From (8-24) and (8-26), we have

$$(2\pi)^4 \int_{-0}^{\infty} dm^2\, \Delta_F(y, m^2)\, \bar{\Delta}(m^2, \varkappa^2) = \pi(\varkappa^2 - y^2 + i\hat{\varepsilon})^{-1} \qquad (8\text{-}27)$$

Therefore if we transform the position-space Feynman integral by $\prod_l \bar{\Delta}(m_l^2, \varkappa_l^2)$, then we obtain $\hat{F}_G(x, \varkappa^2, 1)$ apart from a numerical coefficient, where

$$\hat{F}_G(x, \varkappa^2, \lambda) \equiv \prod_{b \in v(G)-g} \left(\int d^4x_b \right) \prod_{l \in |G|} \{\varkappa_l^2 - \left(\sum_{b \in v(G)} [b:l]\, x_b \right)^2 + i\hat{\varepsilon}_l\}^{-\lambda_l} \qquad (8\text{-}28)$$

The purpose of this subsection is to discuss $\hat{F}_G(x, \varkappa^2, \lambda)$.

It is convenient to replace $\sum_{b \in v(G)} [b:l]\, x_b$ by

$$\sum_{b \in v(G)-g} [b:l]\, x_b + y_l \tag{8-29}$$

where constant position vectors $y_l (l \in |G|)$ are arbitrary. [This arbitrariness is important when we consider nonzero-spin particles.] The original value is obtained by putting

$$y_l = \sum_{b \in g} [b:l]\, x_b \tag{8-30}$$

Now, by means of the generalized Feynman identity (7-1), we can rewrite (8-28) (together with the above-mentioned replacement) as in Subsection 7-1. We use β_l instead of α_l as Feynman parameters. The denominator function (apart from $-i\hat{\varepsilon}$) reads

$$\begin{aligned} \hat{\Phi} &\equiv \sum_{l \in |G|} \beta_l \Big\{ \varkappa_l^2 - \Big(\sum_{b \in v(G)-g} [b:l]\, x_b + y_l \Big)^2 \Big\} \\ &= - \sum_{b,c \in v(G)-g} w_{bc} x_b x_c - 2 \sum_{b \in v(G)-g} \hat{y}_b x_b + \sum_{l \in |G|} \beta_l (\varkappa_l^2 - y_l^2) \end{aligned} \tag{8-31}$$

where

$$\hat{y}_b \equiv \sum_{l \in |G|} [b:l]\, \beta_l y_l \tag{8-32}$$

We can carry out the integrations over $x_b\, (b \in v(G) - g)$ in quite an analogous way to the calculation made in Subsection 7-1. Thus we obtain

$$\begin{aligned} \hat{F}_G(x, \varkappa^2, \lambda) &= (\pi^2 i)^m \, [\Gamma(\lambda_0 - 2m) / \prod_l \Gamma(\lambda_l)] \\ &\quad \times \int_\Delta d^{N-1}\beta \, \frac{\prod_l \beta_l^{\lambda_l - 1}}{\hat{U}^2 (\hat{V} + i\hat{\varepsilon})^{\lambda_0 - 2m}} \end{aligned} \tag{8-33}$$

where $m = M - n$ and n stands for the number of external vertices. The topological formulas for $\hat{U}$ and $\hat{V}$ are found in the following way.

We introduce a tadpole graph G', which is obtained from G by identifying all external vertices; G' has $m + 1$ vertices and N internal lines. We may denote the sole external vertex of G' by g without confusion. The quantities defined with respect to G' are denoted by affixing a prime such as $\boldsymbol{T}'$, $\tilde{U}'$, etc. Hence

$$\tilde{U}' = \sum_{T \in \boldsymbol{T}'} \prod_{l \in T} \beta_l = \sum_{T \in \boldsymbol{T}^n[g]} \prod_{l \in T} \beta_l \tag{8-34}$$

where $T^n[g]$ is the set of all n-trees in G such that any two vertices of g are not linked by the lines of T in G. We may call $\tilde{U}'$ an *n-tree product sum*; it is a homogeneous polynomial of degree m.

Since the matrix $\mathscr{W}'$ of G' can be constructed from $\mathscr{W}$ by summing up all rows of g to make a single row and all columns of g to make a single column, we have

$$\mathscr{W}'^{g} = (w_{bc} \mid b, c \in v(G) - g) \tag{8-35}$$

Hence (8-31) implies

$$\hat{U} = \det \mathscr{W}'^{g} = \tilde{U}' \tag{8-36}$$

with the aid of (3-29). Thus $\hat{U}$ is an n-tree product sum.

THEOREM **8-1** [16] For arbitrary values of y_l, $\hat{V}$ is expressed as follows:

(a) *Cut-set representation*

$$\hat{V} = \sum_{l\in|G|} \beta_l(\varkappa_l^2 - y_l^2) + \tilde{U}'^{-1} \sum_{S\in S'} \tilde{W}_S' \Big(\sum_{l\in|G|} [S:l]\, \beta_l y_l \Big)^2 \tag{8-37}$$

(b) *Circuit representation*

$$\hat{V} = \sum_{l\in|G|} \beta_l \varkappa_l^2 - \tilde{U}'^{-1} \sum_{C\in C'} \tilde{U}_C' \Big(\sum_{l\in|G|} [C:l]\, y_l \Big)^2 \tag{8-38}$$

Proof From (8-31), $\hat{V}$ is defined by

$$\hat{V} \equiv \sum_l \beta_l (\varkappa_l^2 - y_l^2) + \tilde{U}'^{-1} \sum_{b,c\in v(G)-g} [\mathscr{W}'^{g}]^{b,c}\, \hat{y}_b \hat{y}_c \tag{8-39}$$

From Theorems 3-16 and 3-18, we have

$$\begin{aligned} [\mathscr{W}'^{g}]^{b,c} &= \tilde{W}'^{(g|b)} \quad \text{for } b = c \\ &= \tilde{W}'^{(g|bc)} \quad \text{for } b \neq c \end{aligned} \tag{8-40}$$

Hence, with the aid of (3-46), (8-39) is rewritten as

$$\hat{V} = \sum_l \beta_l(\varkappa_l^2 - y_l^2) + \tilde{U}'^{-1} \sum_{u\subset v(G')-g} \tilde{W}'^{(u|u^*)} \Big(\sum_{b\in u} \hat{y}_b \Big)^2 \tag{8-41}$$

where $u^* \equiv v(G') - u$. It is straigthforward to show that (8-41) is equal to (8-37), by using (8-32), (2-10), and (3-44).

To prove (8-38), we make use of Theorem 7-5. Since (7-26) is identically equal to (7-27), we have

$$\begin{aligned} &-\sum_l \alpha_l^{-1} y_l^2 U + \sum_{C\in C} U_C \Big(\sum_l [C:l]\, y_l \Big)^2 \\ &\quad = - \sum_{S\in S} W_S \Big(\sum_l [S:l]\, \alpha_l^{-1} y_l \Big)^2 \end{aligned} \tag{8-42}$$

by setting $q_l = y_l/\alpha_l$ formally. Putting $\alpha_l = 1/\beta_l$ and considering G' instead of G, we obtain

$$-\sum_l \beta_l y_l^2 \tilde{U}' + \sum_{S\in S'} \tilde{W}'_S \Big(\sum_l [S:l]\,\beta_l y_l\Big)^2$$

$$= - \sum_{C\in C'} \tilde{U}'_C \Big(\sum_l [C:l]\,y_l\Big)^2 \tag{8-43}$$

This identity shows that (8-37) equals (8-38). q.e.d.

THEOREM **8-2** [17, 16] In terms of $x_b\,(b\in g)$, $\hat{V}$ has the following *vertex representation*:

$$\hat{V} = \sum_{l\in|G|} \beta_l x_l^2 - \tilde{U}'^{-1} \sum_{\{b,c\}\subset g} \hat{U}_{bc}(x_b - x_c)^2 \tag{8-44}$$

where $\hat{U}_{bc}$ is a sum of $\prod_{l\in T} \beta_l$ over $(n-1)$-trees T in G such that b and c are linked by the lines of T but any vertices of $g - c$ are not linked by them.

Proof We now use (8-30); then

$$\sum_{l\in|G|} [C:l]\,y_l = \sum_{a\in g} \Big(\sum_{l\in|G|} [a:l]\,[C:l]\Big) x_a \tag{8-45}$$

Any circuit C in G' is *either* a circuit *or* a path between two external vertices in G. In the former case, the orthogonality relation (2-8) implies that (8-45) vanishes. In the latter case, if $C\in \boldsymbol{P}_G(bc)$ then we evidently have

$$\sum_l [C:l]\,y_l = \pm(x_b - x_c) \tag{8-46}$$

Thus (8-38) directly implies (8-44) by writing

$$\hat{U}_{bc} \equiv \sum_{C\in P_G(bc)} \tilde{U}'_C \tag{8-47}$$

q.e.d.

§ 9 FEYNMAN-PARAMETRIC FUNCTIONS

9-1 Properties of *U* and *V*

In this subsection, we summarize some properties of U and V as functions of Feynman parameters α_l for fixed values of momenta.

From the definitions, U is a homogeneous polynomial of degree μ and is linear in each α_l, and V is a homogeneous rational function of first order. Furthermore, for all $\alpha_l \geqq 0$, $U \geqq 0$ and U vanishes if and only if $\alpha(C) = 0$ for a circuit C.

THEOREM **9-1** [18] Let H be a connected subgraph of G and $R \equiv G/|H|$. Then

$$U = U_H U_R + U' \tag{9-1}$$

$$V = V_R + V' \tag{9-2}$$

where U_H (or U_R) and V_R denote the chord-set product sum in H (or R) and the V function in R, respectively; U' is a sum of terms whose degree with respect to α_l ($l \in |H|$) is at least $\mu(H) + 1$, and V' is of (at least) first order with respect to α_l ($l \in |H|$).

Proof An arbitrary term of U corresponds to a chord set T^*. Since its complement T is a tree, $T \cap |H|$ includes no circuit, whence Theorem 2-5(d) and the maximality of a tree imply

$$N(T \cap |H|) \leqq M(H) - 1 = N(H) - \mu(H) \tag{9-3}$$

Therefore

$$N(T^* \cap |H|) \geqq \mu(H) \tag{9-4}$$

If the equality does not hold in (9-4), the term corresponding to T^* belongs to U'. Hence we have only to consider the equality case. Then, since $N(T \cap |H|) = M(H) - 1$ and $T \cap |H|$ includes no circuit, $T \cap |H|$ is a tree in H, that is, $T^* \cap |H|$ is a chord set in H. Since

$$N(T \cap |R|) = N(T) - N(T \cap |H|) = M - M(H) = M(R) - 1 \tag{9-5}$$

and since $T \cap |R|$ is connected in R and passes through all vertices of R, $T \cap |R|$ is a tree in R because of Theorem 2-6(c), that is, $T^* \cap |R|$ is a chord set in R. Thus any term of U is contained in the right-hand side of (9-1). Conversely, let T_H and T_R be arbitrary trees in H and R, respectively. Then $T = T_H \cup T_R$ is a tree in G because it cannot include a circuit and $N(T) = M(H) + M(R) - 2 = M - 1$.

Next, we prove (9-2) in the vertex representation (7-45) of V. [Actually, it holds for arbitrary values of q_l, but the proof is more complicated.] We have only to discuss the second term of (7-45). Since $W^{(b|c)}$ is the U function in $G^{(bc)} = G^{(bc)\prime}/l_0$ (see the proof of Theorem 7-7), (9-1) implies

$$W^{(b|c)} = U_H W_R^{(b|c)} + W' \tag{9-6}$$

where $W_R^{(b|c)}$ is the $W^{(b|c)}$ in R and W' is of degree (at least) $\mu(H) + 1$ with respect to $\alpha_l (l \in |H|)$. On substituting (9-6) and (9-1) in (7-45), we easily obtain (9-2) by taking it into account that $W_R^{(b|c)} = 0$ and $W_R^{(a|b)} = W_R^{(a|c)}$ if b and c are not distinct in R. q.e.d.

If $|H|$ consists of only one line, (9-1) reduces to (7-41). If $|H|$ equals a circuit C, (9-1) becomes

$$U = \left(\sum_{l \in C} \alpha_l\right) U_C + U' \tag{9-7}$$

If $|H| = |G| - l$ and if l is not a cut-line in G, then (9-2) reduces to

$$V = \alpha_l m_l^2 + V' \tag{9-8}$$

REMARK **9-2** [18] If we define the value of V at $\alpha(I) = 0$ for any $I \subset |G|$ by $V_{G/I}$, then V is a *continuous* function in $\alpha_l \geqq 0$ for all $l \in |G|$. Therefore V assumes the maximum and the minimum value in the compact region Δ. [As shown by concrete examples with $\mu \geqq 2$, however, $\partial V/\partial \alpha_l$ is bounded but *indefinite* at $\alpha(C) = 0$ if $l \in C \in \mathbf{C}$.] Furthermore, (9-8) implies that if at least one noncut-line l such that $m_l > 0$ exists then V assumes positive values in some subregion of Δ for *any* values of external momenta.

THEOREM **9-3** [3]

$$0 \leqq U \leqq \binom{N}{\mu} (1/N)^{\mu} \quad \text{in } \Delta \tag{9-9}$$

where $\binom{N}{\mu} \equiv N!/\mu!(N - \mu)!$.

Proof From the definition of U we have

$$0 \leqq U \leqq \sum_{I} \prod_{l \in I} \alpha_l \tag{9-10}$$

where the summation goes over all subsets I of $|G|$ such that $N(I) = \mu$. Since the right-hand side of (9-10) is a symmetric expression, its maximum value in Δ is realized at $\alpha_l = 1/N$ for all $l \in |G|$. q.e.d.

A nontrivial lower bound on U will be presented in Subsection 10-1.

9-2 Presence of Nonzero-Spin Particles

As remarked at the end of Subsection 6-2, if $\mathscr{L}_{\text{int}}(x)$ involves nonzero-spin-particle fields and/or derivatives of fields, then the integrand of the Feynman integral has to be multiplied by a certain matrix polynomial, $P(k)$, in $k_l\,(l \in |G|)$. In order to transform the Feynman integral into the Feynman-parametric integral, we have to eliminate the k-dependence of the numerator $P(k)$. For this purpose, it is convenient to introduce the following formal operator [19], called a *D-operator*.

First, we suppose that $P(k)$ is linear in *each* k_l. Then a D-operator D_l is defined by

$$D_l \equiv \frac{1}{2} \int_{m_l^2}^{\infty} dm_l^2 \frac{\partial}{\partial q_l} \tag{9-11}$$

so that

$$k_l(m_l^2 - k_l^2 - i\varepsilon_l)^{-\lambda_l} = D_l(m_l^2 - k_l^2 - i\varepsilon_l)^{-\lambda_l} \tag{9-12}$$

where $k_l = \hat{k}_l + q_l$. In (9-2), $\int_{m_l^2}^{\infty} dm_l^2$ should be understood as an abbreviation of the following two operations: first to integrate the succeeding function over m_l^2 from $\varkappa_l^2$ to infinity and then to put $\varkappa_l = m_l$.

If $P(k)$ is not linear in some k_l, then the definition (9-11) is no longer good because k_l and D_l do not commute. To avoid this difficulty, we define the second D_l to be independent of the first D_l. More precisely, if there is a product $k_l \cdot k_l$ (not a scalar product), we consider

$$D_l^{[2]} D_l^{[1]} \prod_{j=1}^{2} [(m_l^{[j]})^2 - (\hat{k}_l + q_l^{[j]})^2 - i\varepsilon_l]^{-\lambda_l/2} \tag{9-13}$$

with

$$D_l^{[j]} \equiv \frac{1}{2} \int_{m_l^2}^{\infty} d(m_l^{[j]})^2 \frac{\partial}{\partial q_l^{[j]}}\bigg|_{q_l^{[j]}=q_l} \tag{9-14}$$

In the following, we do not explicitly write the superscript of $D_l^{[j]}$.

Since the D-operator involves no integration momenta, the integrations over $\hat{k}_l$ are carried out as in Section 7. Accordingly, we have only to investigate how $P(D)$ operates on (7-24). Since V is the only quantity which depends on m_l and q_l, it is sufficient to know the effect of D_l on V. We define

$$Y_l \equiv -\frac{1}{2\alpha_l} \frac{\partial V}{\partial q_l} \tag{9-15}$$

$$X_{ll'} \equiv -\frac{1}{\alpha_{l'}} \frac{\partial Y_l}{\partial q_{l'}} = X_{l'l} \tag{9-16}$$

Since $\partial X_{ll'}/\partial q_{l''} = 0$, we have (9-17)

$$D_l(V - i\varepsilon)^{-\nu} = Y_l(V - i\varepsilon)^{-\nu}$$

$$D_l D_{l'}(V - i\varepsilon)^{-\nu} = Y_l Y_{l'}(V - i\varepsilon)^{-\nu} + X_{ll'} \frac{(V - i\varepsilon)^{-\nu+1}}{2(-\nu + 1)} \tag{9-18}$$

etc. The general formula is

$$\prod_{i=1}^{k} D_{l_i}(V - i\varepsilon)^{-\nu} = \sum_{m=0}^{[k/2]} \left[\sum_{\text{div}} \left(\prod_{i=1}^{m} X_{l_i' l_i''} \right) \prod_{l \in J_m} Y_l \right] \times \frac{(V - i\varepsilon)^{-\nu+m}}{2^m(-\nu + 1) \ldots (-\nu + m)} \tag{9-19}$$

where $[k/2]$ stands for the greatest integer not exceeding $k/2$, and the summation $\sum_{\text{div}}$ goes over all possible divisions of the set $\{l_1, \ldots, l_k\}$ into $m + 1$ disjoint subsets $\{l_1', l_1''\}, \ldots, \{l_m', l_m''\}$, and J_m.

From the circuit representation (7-26) and the cut-set one (7-27) of V, the following expressions for Y_l and $X_{ll'}$ are immediately obtained:

$$Y_l = q_l - U^{-1} \sum_{C \in \mathbf{C}} [C : l]\, U_C \Big(\sum_{l \in |G|} [C : l]\, \alpha_l q_l \Big) \tag{9-20}$$

$$Y_l = U^{-1} \sum_{S \in \mathbf{S}} [S : l]\, (W_S/\alpha_l) \Big(\sum_{l \in |G|} [S : l]\, q_l \Big) \tag{9-21}$$

$$X_{ll'} = U^{-1} \sum_{C \in \mathbf{C}} [C : l]\, [C : l']\, U_C \delta \tag{9-22}$$

$$X_{ll'} = -U^{-1} \sum_{S \in \mathbf{S}} [S : l]\, [S : l']\, (W_S/\alpha_l \alpha_{l'})\, \delta \tag{9-23}$$

where δ denotes the metric tensor of the Minkowski space (see Subsection 6-1). In particular, for $l' = l$ (9-22) reduces to

$$X_{ll} = U^{-1}\, \partial U/\partial x_l\, \delta \tag{9-24}$$

with the aid of (7-33).

THEOREM **9-4** [3] In terms of external momenta, Y_l has the following *vertex representation*:

$$Y_l = U^{-1} \sum_{b \in v(G) - a} W_l^{(a|b)} p_b \tag{9-25}$$

with

$$W_l^{(a|b)} \equiv \sum_{P \in \mathbf{P}(ab)} [P : l]\, U_P \tag{9-26}$$

where U_P is the U function in G/P (see Theorem 3-20), and $[P : l] = 0$ for $l \notin P$ and $[P : l] = \pm 1$ for $l \in P$ (it is plus if l has the same orientation as P and minus if opposite, where P is supposed to be oriented from a to b).

Proof Since external momenta p_b $(b \neq a)$ can be independently assigned, it is sufficient to prove

$$Y_l = U^{-1} \sum_{P \in \boldsymbol{P}(ab)} [P:l]\, U_P\, p_b \tag{9-27}$$

in the case of two external vertices a and b $(p_a + p_b = 0)$.

From (9-21) we have

$$Y_l = U^{-1} \sum_{S \in \boldsymbol{S}(a|b)} [S:l]\,(W_S/\alpha_l)\, p_b \tag{9-28}$$

where the overall sign of $[S:l]$ is fixed by putting $a_S = a$ (see Definition 2-12). Therefore, we have only to prove

$$\sum_{S \in \boldsymbol{S}(a|b)} [S:l]\,\beta_l \sum_{T^2 \in \boldsymbol{T}^2(S)} \prod_{k \in T^2} \beta_k$$

$$= \sum_{P \in \boldsymbol{P}(ab)} [P:l] \sum_{T \supset P} \prod_{k \in T} \beta_k \tag{9-29}$$

Given a 2-tree $T^2 \in \boldsymbol{T}^2(S)$ with $l \in S \in \boldsymbol{S}(a \mid b)$, $T \equiv T^2 + l$ is evidently a tree, and therefore it includes a unique path P between a and b; P contains l because a and b are not linked by $T - l$. Conversely, given a tree T which includes a path P between a and b such that $l \in P$, $T^2 \equiv T - l$ is a 2-tree by definition, and there exists a unique cut-set S such that $l \in S$ and $S \cap T^2 = \emptyset$ because of Theorem 2-29. Since $P \subset T$ and $S \cap (T - l) = \emptyset$, we have $S \cap P = \{l\}$, and therefore $S \in \boldsymbol{S}(a \mid b)$ because if both a and b are involved in one connected subgraph of $G - S$ then $N(S \cap P)$ has to be even [cf. (2-11)]. Furthermore, since $S \cap P = \{l\}$, we find

$$[S:l] = [P:l] \tag{9-30}$$

as seen by considering $G/(|G| - S)$. q.e.d.

In particular, if l is a cut-line, then (9-27) reduces to Theorem 3-20, because all indices $[P:l]$ for $P \in \boldsymbol{P}(ab)$ are mutually equal and (9-20) reduces to $Y_l = q_l = [P:l]\,p_b$. Nontrivial examples of Theorem 9-4 will be presented in Subsection 12-1.

THEOREM **9-5** [3] The identity

$$\partial V/\partial \alpha_l = m_l^2 - Y_l^2 \tag{9-31}$$

holds for arbitrary values of q_l.

Proof From (7-50)

$$V = \sum_k \alpha_k m_k^2 - U^{-1} \det \mathscr{V}_{T^*} \tag{9-32}$$

where $\mathscr{V}_{T^*}$ is defined by (7-49) and $T^* \equiv \{l_1, \ldots, l_\mu\}$ is a chord set. Suppose $l = l_1$; then

$$\frac{\partial V}{\partial \alpha_l} = m_l^2 - U^{-1} \frac{\partial}{\partial \alpha_l} \det \mathscr{V}_{T^*} + U^{-2} \frac{\partial U}{\partial \alpha_l} \det \mathscr{V}_{T^*} \tag{9-33}$$

with

$$(\partial/\partial\alpha_l) \det \mathscr{V}_{T^*} = q_l^2 U + 2q_l[\mathscr{V}_{T^*}]^{0,1} + [\mathscr{V}_{T^*}]^{1,1} \tag{9-34}$$

where $[\mathscr{V}_{T^*}]^{i,j}$ denotes the (i, j) cofactor of $\mathscr{V}_{T^*}$, whose rows and columns are numbered from 0 to μ. Since $[\mathscr{V}_{T^*}]^{0,0} = \det \mathscr{V}_{T^*} = U$ and $[\mathscr{V}_{T^*}]^{01,01} = \partial U/\partial\alpha_l$, where $[\mathscr{V}_{T^*}]^{01,01}$ is a principal minor of $\mathscr{V}_{T^*}$ obtained by deleting the first two rows and two columns, Jacobi's theorem for determinants yields

$$\frac{\partial U}{\partial \alpha_l} \det \mathscr{V}_{T^*} = [\mathscr{V}_{T^*}]^{01,01} \det \mathscr{V}_{T^*} = \begin{vmatrix} [\mathscr{V}_{T^*}]^{0,0} & [\mathscr{V}_{T^*}]^{0,1} \\ [\mathscr{V}_{T^*}]^{1,0} & [\mathscr{V}_{T^*}]^{1,1} \end{vmatrix}$$

$$= U[\mathscr{V}_{T^*}]^{1,1} - ([\mathscr{V}_{T^*}]^{0,1})^2 \tag{9-35}$$

On substituting (9-34) and (9-35) in (9-33), we find

$$\partial V/\partial\alpha_l = m_l^2 - (q_l + U^{-1} [\mathscr{V}_{T^*}]^{0,1})^2 = m_l^2 - Y_l^2 \tag{9-36}$$

q.e.d.

Note The position-space Feynman-parametric integral (8-33) can similarly be modified when nonzero-spin particles are present. In this case, the numerator is a certain matrix function of $\partial/\partial\beta_l\,(l \in |G|)$, whence it is unnecessary to introduce any special operator such as D_l.

§ 10 ULTRAVIOLET DIVERGENCE

10-1 Proof of Dyson's Power-Counting Theorem

As remarked in Subsection 6-2, the Feynman integral

$$P(D)\,F_G(p) = \sum_{l\in|G|}\left(\int d^4k_l\right)\frac{\prod_{b\in v(G)}\delta^4\left(\sum_l [b:l]\,k_l + p_b\right)}{\prod_{l\in|G|}(m_l^2 - k_l^2 - i\varepsilon_l)}\,P(k) \qquad (10\text{-}1)$$

is often ultraviolet divergent. The convergence criterion for it is known as Dyson's power-counting theorem [20]: (10-1) is free from ultraviolet divergence if the inequality

$$2N\left(\bigcup_{i=1}^{j} C_i\right) - 4j - \nu\left(\bigcup_{i=1}^{j} C_i\right) > 0 \qquad (10\text{-}2)$$

holds for any nonemtpy set, $\{C_1, \ldots, C_j\}$, of independent circuits, where $\nu(I)$ denotes the degree of $P(k)$ with respect to $k_l\,(l \in I)$. Because of the indefinite character of (10-1), Dyson's reasoning cannot be regarded as a proof. The purpose of this subsection is to present a rigorous proof of the above statement on the basis of the Feynman-parametric integral.

THEOREM **10-1** [21] Let

$$\sigma \equiv \max_I \mu(I)/N(I) \qquad (10\text{-}3)$$

where I is any nonempty union of circuits. Then

$$U \geqq \prod_{l\in|G|} \alpha_l^\sigma \quad \text{in } \Delta \qquad (10\text{-}4)$$

Proof We number all internal lines as $1, \ldots, N$. We may assume

$$1 \geqq \alpha_1 \geqq \alpha_2 \geqq \ldots \geqq \alpha_N \geqq 0 \qquad (10\text{-}5)$$

without loss of generality. We introduce a linear (lexicographic) ordering among chord sets in the following way. Let $T^{(k)*} = \{l_1^{(k)}, \ldots, l_\mu^{(k)}\}$ with $l_1^{(k)} < \ldots < l_\mu^{(k)}$; then we write $T^{(1)*} < T^{(2)*}$ if and only if there exists a number $j\,(1 \leqq j \leqq \mu)$ such that $l_1^{(1)} = l_1^{(2)}, \ldots, l_{j-1}^{(1)} = l_{j-1}^{(2)}$, and $l_j^{(1)} < l_j^{(2)}$. We denote by T^* the *first* chord set with respect to the above ordering. Let $T^* = \{l_1, \ldots, l_\mu\}$ with $l_1 < \ldots < l_\mu$. By the definition of U, we have

$$U \geqq \prod_{j=1}^{\mu} \alpha_{l_j} \qquad (10\text{-}6)$$

Let C_j be the circuit for which $T^* \cap C_j = \{l_j\}$ (see Theorem 2-22), and put

$$I_j \equiv \bigcup_{k=j}^{\mu} C_k \quad (j = 1, 2, \ldots, \mu) \tag{10-7}$$

with $I_{\mu+1} \equiv \emptyset$. Then of course

$$\mu(I_j) = \mu(I_{j+1}) + 1 \quad (j = 1, \ldots, \mu) \tag{10-8}$$

Furthermore, (10-3) implies

$$\mu(I_j) \leqq \sigma N(I_j) \quad (j = 1, \ldots, \mu) \tag{10-9}$$

Since $1 \geqq \alpha_{l_1} \geqq \ldots \geqq \alpha_{l_\mu} \geqq 0$, therefore, writing $\alpha_{l_0} \equiv 1$ we have

$$\prod_{j=1}^{\mu} \alpha_{l_j} = \prod_{j=1}^{\mu} (\alpha_{l_j}/\alpha_{l_{j-1}})^{\mu(I_j)} \geqq \left(\prod_{j=1}^{\mu} \alpha_{l_j}^{N_j} \right)^{\sigma} \tag{10-10}$$

where

$$N_j \equiv N(I_j) - N(I_{j+1}) \quad (j = 1, \ldots, \mu) \tag{10-11}$$

Finally, we note

$$\alpha_l \leqq \alpha_{l_j} \quad \text{for any} \quad l \in I_j \tag{10-12}$$

because if $\alpha_l > \alpha_{l_j}$ for some $l \in I_j$ then $l < l_j$, whence there would exist a chord set T'^* such that $T'^* < T^*$ in contradiction with the definition, where T'^* is defined to be a union of $\{l_1, \ldots, l_{j-1}\}$ and a chord set in $G/(|G| - I_j)$ containing l [see the proof of (9-1)]. Therefore

$$\prod_{j=1}^{\mu} \alpha_{l_j}^{N_j} \geqq \prod_{j=1}^{\mu} \prod_{l \in I_j - I_{j+1}} \alpha_l = \prod_{l \in I_1} \alpha_l \geqq \prod_{l \in |G|} \alpha_l \tag{10-13}$$

Collecting (10-6), (10-10), and (10-13), we obtain (10-4). q.e.d.

Theorem 10-2 When momenta are fixed, Y_l is bounded in Δ.

Proof From (7-34), for any $S \in \boldsymbol{S}$ we have

$$\alpha_l U \geqq \sum_{S' \ni l} W_{S'} \geqq |[S:l]|\, W_S \tag{10-14}$$

Hence (9-21) implies

$$|Y_l^{(\varrho)}| \leqq \sum_{S \in \boldsymbol{S}} \left| \sum_{l \in |G|} [S:l]\, q_l^{(\varrho)} \right| \quad (\varrho = 0, 1, 2, 3) \tag{10-15}$$

where (ϱ) is the 4-vector-component index. q.e.d.

Theorem 10-3 The singularities of $X_{ll'}$ in Δ are first-order poles at $\alpha(C) = 0$, where C is any circuit containing both l and l'. More precisely,

$$|X_{ll'}^{(\varrho\varrho)}| \leqq \sum_{C \ni l, l'} \left(\sum_{k \in C} \alpha_k \right)^{-1} \quad \text{in } \Delta \quad (\varrho = 0, 1, 2, 3) \tag{10-16}$$

Proof From (9-7) we find

$$\Big(\sum_{k\in C} \alpha_k\Big) U_C \leqq U \tag{10-17}$$

Therefore (9-22) yields (10-16). q.e.d.

THEOREM **10-4** When ε is finite, the Feynman parametric integral $P(D) f_G(p)$ is absolutely convergent if

$$2N(I) - 4\mu(I) - \nu(I) > 0 \tag{10-18}$$

for any nonempty union I of circuits, where $\nu(I)$ denotes the degree of $P(D)$ with respect to D_l $(l \in I)$.

Proof Because of (9-19), we have only to show the convergence of the integral

$$\int_\Delta d^{N-1}\alpha \frac{\prod_{i=1}^{m} Xl_{l_i' l_i''} \prod_{l\in J_m} Y_l}{U^2(V - i\varepsilon)^{N-2\mu-m}} \tag{10-19}$$

where the definition of $\nu(I)$ implies

$$N(\{l_1, \ldots, l_k\} \cap I) \leqq \nu(I) \tag{10-20}$$

In particular, $m \leqq \nu(|G|)/2$, whence (10-18) assures $N - 2\mu - m > 0$. Since ε is finite, Theorems 10-2 and 10-3 imply that (10-19) is absolutely convergent if the integral

$$B \equiv \int_\Delta \frac{d^{N-1}\alpha}{U^2 \prod_{i=1}^{m} \Big(\sum_{k\in C_i} \alpha_k\Big)} \tag{10-21}$$

is convergent, where C_i is any circuit containing l_i' and l_i''. Let

$$2\sigma' \equiv \max_I \frac{2\mu(I) + [\nu(I)/2]}{N(I)} \tag{10-22}$$

Then (10-18) implies $2\sigma' < 1$. Therefore if we can prove

$$U^2 \prod_{i=1}^{m} \Big(\sum_{k\in C_i} \alpha_k\Big) \geqq \prod_{l\in|G|} \alpha_l^{2\sigma'} \tag{10-23}$$

then (7-1) leads us to

$$0 < B < [\Gamma(1 - 2\sigma')]^N / \Gamma(N(1 - 2\sigma')) < \infty \tag{10-24}$$

The proof of (10-23) is quite analogous to that of Theorem 10-1.

Suppose that $T^* = \{l_1, \ldots, l_\mu\}$ and $I_1, \ldots, I_{\mu+1}$ have the same meanings as in the proof of Theorem 10-1. Given an arbitrary circuit C, we can always find j such that $C \subset I_j$ but $C \not\subset I_{j+1}$. Then since $C \cup I_{j+1} = I_j$ (because $C \cup I_{j+1} \subset I_j$ and both are unions of $\mu - j + 1$ independent circuits), for α_{l_j} satisfying (10-12) we have

$$\sum_{k \in C} \alpha_k \geqq \alpha_{l_j} \tag{10-25}$$

because $l_j \in C$.

The number n_j of the circuits included in I_j among $C_1, \ldots, C_m$ is, by definition, not greater than

$$[N(\{l_1', l_1'', \ldots, l_m', l_m''\} \cap I_j)/2] \tag{10-26}$$

which is, in turn, not greater than $[\nu(I_j)/2]$ because of (10-20). Hence, with the aid of (10-25), we have

$$\begin{aligned} \prod_{i=1}^{m} \Big(\sum_{k \in C_i} \alpha_k \Big) &\geqq \prod_{j=1}^{\mu} \alpha_{l_j}^{n_j - n_{j+1}} \\ &= \prod_{j=1}^{\mu} (\alpha_{l_j}/\alpha_{l_{j-1}})^{n_j} \\ &\geqq \prod_{j=1}^{\mu} (\alpha_{l_j}/\alpha_{l_{j-1}})^{[\nu(I_j)/2]} \end{aligned} \tag{10-27}$$

Combining (10-27) with (10-6) and using (10-22) and (10-13), we find

$$\begin{aligned} U^2 \prod_{i=1}^{m} \Big(\sum_{k \in C_i} \alpha_k \Big) &\geqq \prod_{j=1}^{\mu} (\alpha_{l_j}/\alpha_{l_{j-1}})^{2\mu(I_j) + [\nu(I_j)/2]} \\ &\geqq \Big(\prod_{j=1}^{\mu} \alpha_{l_j}^{N_j} \Big)^{2\sigma'} \\ &\geqq \Big(\prod_{l \in |G|} \alpha_l \Big)^{2\sigma'} \end{aligned} \tag{10-28}$$

with N_j defined by (10-11). q.e.d.

The condition (10-18) is evidently equivalent to (10-2).

10-2 Renormalization

In order to avoid ultraviolet divergence, the Feynman integral (10-1) has to be modified. One of such modifications is to consider $P(D)\, F_G(p, \lambda)$, but this replacement is not preferable from the physical point of view because its value is much different from that of $P(D)\,F_G(p)$ when it converges. The most straightforward practical way is to replace the propagator (6-15) by

$$\frac{1}{m_l^2 - k_l^2 - i\varepsilon_l}\left[\frac{\Lambda_l^2 - m_l^2}{\Lambda_l^2 - k_l^2 - i\varepsilon_l}\right]^{\lambda_l} = \lambda_l \int_{m_l^2}^{\Lambda_l^2} d\varkappa_l^2 \, \frac{(\varkappa_l^2 - m_l^2)^{\lambda_l - 1}}{(\varkappa_l^2 - k_l^2 - i\varepsilon_l)^{\lambda_l + 1}} \tag{10-29}$$

where $\Lambda_l > m_l$ and $\lambda_l \geqq 0$; (10-29) tends to (6-15) as $\lambda_l \to 0$, and it also does as $\Lambda_l \to \infty$. The Feynman integral cut off in this way converges for appropriate values of λ_l ($l \in |G|$), and it will not be appreciably different from the original value (when finite) if any of Λ_l ($l \in |G|$) is much larger than all masses and all external momenta in their magnitudes. The case $\lambda_l = 1$ is called the *Feynman cutoff* [1].

The above cutoff method is simple, but it destroys the unitarity condition (7-65). The most satisfactory way of eliminating ultraviolet divergence is known as *renormalization*. We formally introduce some additional terms containing infinite constants into the interaction Lagrangian in such a way that the ultraviolet divergence of every Feynman integral is canceled out. In certain theories, called *renormalizable field theories*, the number of the infinite constants is finite, and they can be interpreted as changes of masses and coupling constants, which are called *mass* and *coupling-constant renormalizations*. It needs a lot of analysis to show the possibility of renormalization, and since it is described in many standard textbooks on quantum field theory [10, 11, 12], we do not repeat it here.

When we are concerned with a Feynman integral corresponding to G, renormalization implies to subtract some Feynman integrals corresponding to reduced graphs of G multiplied by infinite constants. A rigorous formulation of renormalizing the Feynman integral was introduced by Bogoliubov and Parasiuk [22] and completed by Hepp [23]. We here briefly describe two other formulations in which the renormalized Feynman integral is written in closed form.

The first one is due to Speer [24], who generalized the definition of the finite part due to Riesz [25]. Given a divergent integral B, its *finite part* can be defined consistently in the following way. We consider an integral $B(\lambda)$ dependent on a parameter λ such that $B(\lambda)$ is convergent in a certain domain D of λ and it formally coincides with B at $\lambda = \lambda_0$. If the analytic continuation of $B(\lambda)$ from D uniquely has a Laurent expansion at $\lambda = \lambda_0$, then the finite part of B is defined to be its constant term.

The λ-dependent Feynman integral $P(D)\, f_G(p, \lambda)$ is certainly free from ultraviolet divergence, and therefore holomorphic in λ, if

$$2 \sum_{l \in I} \operatorname{Re} \lambda_l > 4\mu(I) + \nu(I) \tag{10-30}$$

for any non-empty union I of circuits. Furthermore, we can prove that $P(D)\, f_G(p, \lambda)$ is a meromorphic function of $\lambda_l\,(l \in |G|)$ for $\operatorname{Re} \lambda_l > -L$ with any $L > 0$; its poles are located only at the points at which for some $I \subset |G|$ $\sum_{l \in I} \lambda_l$ becomes an integer.

The finite part should be defined in such a way that it satisfies certain natural requirements. The finite part, $\bar{f}_G(p)$, of $f_G(p)$ can be defined to be the nonsingular part of $f_G(p, \lambda)$ at $\lambda = 1$ given by

$$\operatorname*{Reg}_{\lambda = 1} f_G(p, \lambda) \equiv (2\pi i)^{-N} \operatorname{Sym} \prod_{l \in |G|} \left(\int_{\Gamma_l} d\lambda_l \right) f_G(p, \lambda) / \prod_{l \in |G|} (\lambda_l - 1) \tag{10-31}$$

where Γ_l is a small circle $|\lambda_l - 1| = r_l$ oriented counterclockwise such that $r_l \neq r_{l'}$ for any $l \neq l'$ and $r_l > r_{l'}$ implies $r_l > 2r_{l'}$, and the symbol Sym means to take the arithmetic mean over all permutations of the relative magnitudes of r_l $(l \in |G|)$.

The above prescription is not only equivalent to the conventional renormalization procedure apart from a finite renormalization in the renormalizable field theories but also applicable to the unrenormalizable field theories. Unfortunately, however, one cannot regard (10-31) as a physically significant method of calculating all ultraviolet-divergent quantities without cutoff.

Example For the $N = 2$ self-energy graph G (the two internal-line particles are spinless and have mass m), we have

$$f_G(p; \lambda_1, \lambda_2) = \frac{\Gamma(\lambda_1 + \lambda_2 - 2)}{\Gamma(\lambda_1)\,\Gamma(\lambda_2)} \int_0^1 d\alpha \, \frac{\alpha^{\lambda_1 - 1}\,(1 - \alpha)^{\lambda_2 - 1}}{[m^2 - \alpha(1 - \alpha)\, p^2 - i\varepsilon]^{\lambda_1 + \lambda_2 - 2}}$$

Because of the symmetry of the integrand, we may first carry out the contour integration over λ_2 which is equivalent to set $\lambda_2 = 1$. Then

$$f_G(p; \lambda, 1) = (\lambda - 1)^{-1} \int_0^1 d\alpha \, \alpha^{\lambda - 1} \{[m^2 - \alpha(1 - \alpha)\, p^2 - i\varepsilon]^{-\lambda + 1} - 1\}$$
$$+ (\lambda - 1)^{-1} \lambda^{-1}$$

Since the first term is holomorphic at $\lambda = 1$, we find

$$\bar{f}_G(p) = -\int_0^1 d\alpha \log [m^2 - \alpha(1-\alpha)p^2 - i\varepsilon] - 1$$

Thus the result is dependent on the choice of mass unit.

The second method of obtaining the renormalized Feynman-parametric integral is due to Lam [26] and Appelquist [27]. It is more closely along the line of the conventional renormalization theory [20] and based on the formula for the remainder of a Taylor series:

$$F(x_1, \ldots, x_k) - \sum_{j=0}^{n} \frac{1}{j!}\left(\frac{\partial}{\partial \xi}\right)^j F(\xi x_1, \ldots, \xi x_k)\bigg|_{\xi=0} = \int_0^1 d\xi \, \frac{(1-\xi)^n}{n!}\left(\frac{\partial}{\partial \xi}\right)^{n+1} F(\xi x_1, \ldots, \xi x_k) \tag{10-32}$$

Let I be a nonempty union of circuits such that

$$d(I) \equiv 4\mu(I) + \nu(I) - 2N(I) \geqq 0 \tag{10-33}$$

and H be a sub-Feynman-graph such that $|H| = I$. From the sub-Feynman-integral corresponding to H we subtract its Taylor expansion to order $d(I)$ around the point at which all external momenta of H are zero, and make use of (10-32) (we denote the integration parameter by ξ_I). After doing this for all possible circuit-unions I such that $d(I) \geqq 0$ and H is nonseparable, we convert the subtracted Feynman integral into the Feynman-parametric form. The resulting expression becomes particularly simple when all particles are spinless ($\nu(I) = 0$). By putting $\xi_I^2 = \zeta_I$, the renormalized Feynman-parametric integral explicitly reads [cf. (7-71)]

$$\prod_l \left(\int_0^\infty d\alpha_l\right) \prod_I \left[\int_0^1 d\zeta_I \frac{(1-\zeta_I)^{d_I}}{d_I!}\left(\frac{\partial}{\partial \zeta_I}\right)^{d_I+1}\right] \frac{i^{N-2\mu}}{\bar{U}^2} e^{-i\bar{V}-\varepsilon} \tag{10-34}$$

where $d_I \equiv 2\mu(I) - N(I) \geqq 0$, and $\bar{U}$ and $\bar{V}$ are defined as follows:

$$\bar{U} \equiv \sum_{T^* \in \mathbf{T}^*} \prod_{l \in T^*} \Big(\alpha_l \prod_{I \ni l} \zeta_I\Big) \Big/ \prod_I \zeta_I^{\mu(I)} \tag{10-35}$$

$$\bar{V} \equiv \sum_{l \in |G|} \alpha_l m_l^2 + \bar{U}^{-1} \sum_{\{b,c\} \subset g} \bar{W}^{(b|c)} p_b p_c \tag{10-36}$$

with $\bar{W}^{(b|c)}$ being the $\bar{U}$ function in $G^{(bc)}$. As seen from (9-1), $\bar{U}$ is a polynomial not only in α_l but also in ζ_I; when $\zeta_I = 0$ and $\zeta_J = 1$ for all $J \neq I$, $\bar{U}$ reduces to $U_H U_R$, where $|H| = I$ and $R = G/I$.

In order to see the meaning of (10-34), we consider the simplest case in which only one I is involved in (10-34) and $d_I = 0$. Then we have

$$\prod_{l \in |G|} \left(\int_0^\infty d\alpha_l \right) \int_0^1 d\zeta_I \frac{\partial}{\partial \zeta_I} \frac{\exp(-i\bar{V} - \varepsilon)}{\bar{U}^2}$$

$$= \lim_{\eta \to +0} \left[\prod_{l \in |G|} \left(\int_\eta^\infty d\alpha_l \right) \frac{\exp(-iV - \varepsilon)}{U^2} \right.$$

$$- \prod_{l \in |H|} \left(\int_\eta^\infty d\alpha_l \right) \frac{\exp[-iV_H(0) - \varepsilon_H]}{U_H^2}$$

$$\left. \times \prod_{l \in |R|} \left(\int_\eta^\infty d\alpha_l \right) \frac{\exp(-iV_R - \varepsilon_R)}{U_R^2} \right] \tag{10-37}$$

with the aid of Theorem 9-1, where $V_H(0) \equiv \sum_{l \in |H|} \alpha_l m_l^2$ and $\varepsilon_H + \varepsilon_R = \varepsilon (\varepsilon_H \geqq 0, \varepsilon_R \geqq 0)$. The second term of (10-37) is the subtraction term, which is an infinite constant times the Feynman-parametric integral corresponding to R.

References

(1) R. P. Feynman, *Phys. Rev.* **76**, 769 (1949).
(2) R. Chisholm, *Proc. Camb. Phil. Soc.* **48**, 300 (1952); Erratum, **48**, 518 (1952).
(3) N. Nakanishi, *Progr. Theoret. Phys.* **17**, 401 (1957).
(4) Y. Nambu, *Nuovo Cim.* **6**, 1064 (1957).
(5) K. Symanzik, *Progr. Theoret. Phys.* **20**, 690 (1958).
(6) N. Nakanishi, *Progr. Theoret. Phys.* **26**, 337 (1961).
(7) Y. Shimamoto, *Nuovo Cim.* **25**, 1292 (1962).
(8) T. Kinoshita, *J. Math. Phys.* **3**, 650 (1962).
(9) R. Jost, *General Theory of Quantized Fields* (American Math. Soc. Publications, 1963).
(10) H. Umezawa, *Quantum Field Theory* (North Holland, Amsterdam, 1956).
(11) S. S. Schweber, *An Introduction to Quantum Field Theory* (Harper & Row, New York, 1961).

(12) J. D. Bjorken and S. D. Drell, *Relativistic Quantum Fields* (McGraw-Hill, New York, 1965).
(13) J. Kahane, *J. Math. Phys.* **9**, 1732 (1968).
(14) N. Nakanashi, *Progr. Theoret. Phys. Suppl.* **18**, 1 (1961); Errata, *Progr. Theoret. Phys.* **26**, 806 (1961).
(15) T. T. Wu, *Phys. Rev.* **123**, 678 (1961).
(16) N. Nakanashi, *Progr. Theoret. Phys.* **42**, 966 (1969).
(17) C. S. Lam and J. P. Lebrun, *Nuovo Cim.* **59**A, 397 (1969).
(18) N. Nakanishi, *Progr. Theoret. Phys.* **22**, 128 (1959).
(19) R. Karplus and N. M. Kroll, *Phys. Rev.* **77**, 536 (1950).
(20) F. J. Dyson, *Phys. Rev.* **75**, 1736 (1949).
(21) N. Nakanishi, *J. Math. Phys.* **4**, 1385 (1963).
(22) N. N. Bogoliubov and O. S. Parasuik, *Acta Math.* **97**, 227 (1957).
(23) K. Hepp, *Commun. Math. Phys.* **2**, 301 (1966).
(24) E. R. Speer, *J. Math. Phys.* **9**, 1404 (1968).
(25) M. Riesz, *Acta Math.* **81**, 1 (1949).
(26) C. S. Lam, *Nuovo Cim.* **59**A, 422 (1969).
(27) T. Appelquist, *Ann. of Phys.* **54**, 27 (1969).

CHAPTER 3

Singularities of the Feynman Integral

The Feynman integral can be regarded as a boundary value of an analytic function of complex momenta, which is called a Feynman function. It is a very interesting and important problem to investigate its analytic properties, that is, to find the locations of its singularities and their natures. Necessary conditions for the existence of a singularity are expressed in terms of certain equations for momenta and Feynman parameters, which are usually called Landau equations. Some general properties of the singularity can be discussed by means of the Landau equations together with the Feynman-parametric formula. More detailed study of the singularity structures is, however, extremely difficult for the general Feynman graph; only some low-order or special-type graphs can be analyzed in details.

In 1951, Eden [1] first discussed the threshold behavior of a Feynman integral as an analytic function of an energy variable. In 1958, the existence of anomalous thresholds, which could not be expected from the unitarity of the S-matrix, was discovered by Karplus, Sommerfield, and Wichmann [2], by Nambu [3], and by Oehme [4], independently. In 1959, systematic investigations of the singularities of the general Feynman integral were made by Landau [5], by Nakanishi [6], and by Bjorken [7], independently. Since then, many authors have investigated detailed analytic properties for various Feynman graphs. Especially, people of the Cambridge school [8] extensively developed a general but heuristic theory on the analyticity of the Feynman function. Cutkosky [9] proposed an interesting rule for giving the discontinuity around a singularity of the Feynman function. In 1964, Fotiadi, Froissart, Lascoux, and Pham [10] introduced the homological method, which is based on rigorous mathematics and is powerful in investigating the Riemann-sheet structure of the Feynman function.

In this chapter, we develop the theory of the singularities of the general Feynman function on the basis of the Feynman-parametric formula. In Section 11, the Feynman function is defined, but we do not discuss the homological method in detail. In Section 12, we derive the Landau equations and discuss some general properties of the Landau singularities. Section 13 is devoted to the investigation of real singularities such as normal and anomalous thresholds. Physical-region singularities and infrared divergence are discussed in Section 14. In Section 15, we investigate the region swept by the singularities as internal masses are varied. Finally, the second-type singularities are discussed in Section 16.

We do not claim mathematical rigor throughout this chapter.

§ 11 FEYNMAN FUNCTIONS

It is the purpose of this chapter to investigate how the Feynman integral behaves when external momenta p_b ($b \in g$) vary. Throughout this chapter, $n(\geqq 2)$ stands for the number of the external vertices of a connected Feynman graph G. The collection of $n - 1$ (independent) external momenta is symbolically denoted by p.

The renormalized Feynman integral $\bar{f}_G(p)$ with $\varepsilon \to +0$ is a distribution over external momenta p. According to a well-known result in the axiomatic field theory [11], the extended transition amplitude is a certain boundary value of an analytic function of external momenta. Because of Theorem 7–8, therefore, the Feynman integral should also be a boundary value of an analytic function. This analytic function is called a *Feynman function* and denoted by $\bar{f}_G(p)$. It is an analytic function of $n - 1$ complex 4-momenta.

Given a Feynman integral, it is a boundary value of an analytic function of $n - 1$ 4-momenta as mentioned above. One should note, however, that it may not necessarily be a boundary value of an analytic function if some constraints are imposed on external momenta, for example, if they are fixed on the mass shells.

Since $\bar{f}_G(p)$ is Lorentz-invariant, it is sometimes more convenient to regard it as a function of invariant variables. According to a theorem due to Hall and Wightman [12], the Feynman function $\bar{f}_G(p)$ is an analytic function of independent scalar products of external momenta. We regard it as a function of independent ones of invariant squares

$$s_h \equiv \Big(\sum_{b\in h} p_b\Big)^2 = \Big(\sum_{b\in g-h} p_b\Big)^2 \tag{11-1}$$

for subsets h of g, because they are physically more convenient quantities than scalar products.

When nonzero-spin particles are taken into account, the Feynman function of course can be introduced in exactly the same way as above, but it is not Lorentz-invariant but covariant. It can be written as a linear combination of Lorentz-invariant functions, each of which is a function of invariant squares (11-1). The analytic properties of the Feynman function is determined to a large extent by the denominator function V of the Feynman-parametric formula, and the modifications due to the presence of spins bring only minor changes. Hence, we are mainly concerned with the theory of spinless particles alone.

Let $a \in g$ and

$$D_V \equiv \{p_b\,(b \in g - a) \mid V(p, \alpha) \neq 0 \quad \text{for all } \alpha \in \Delta\} \tag{11-2}$$

where $V(p, \alpha)$ is given by (7-45) or (7-46) together with $\sum_{b \in g} p_b = 0$ and $\alpha \in \Delta$ means $\alpha_l \geqq 0$ for all $l \in |G|$ and $\sum_l \alpha_l = 1$. Then the integral $\operatorname{Reg}_{\lambda=1} f_G(p, \lambda)$ with

$$f_G(p, \lambda) \equiv \frac{\Gamma(\lambda_0 - 2\mu)}{\prod_l \Gamma(\lambda_l)} \int_\Delta d^{N-1}\alpha \, \frac{\prod_l \alpha_l^{\lambda_l - 1}}{U^2(\alpha)\,[V(p, \alpha)]^{\lambda_0 - 2\mu}} \tag{11-3}$$

for p complex defines a holomorphic function of p in each connected part of D_V. When it is analytically continued, it defines the Feynman function $\bar{f}_G(p)$, which is, in general, not single-valued. The Riemann sheet on which the boundary value yields the Feynman integral is called the *physical sheet.*

The Riemann-sheet structure of the Feynman function can be investigated by the *homological method* [13, 14], which is an elegant application of modern mathematics. In the homological method, one usually starts from the momentum-space Feynman integral (6-26) without $\delta^4(\sum p_b)$, where the indices λ_l ($l \in |G|$) are chosen to be positive integers. The Feynman function can be defined from it by dropping $-i\varepsilon_l$ ($l \in |G|$) and regarding it as a multiple contour integral. This definition is, however, somewhat ambiguous because contours are not compact. Therefore, one has to compactify the momentum integrations before applying the homological theory. The difficulty remarked in (2) of Subsection 6-2 is inherited by the ambiguity of the method of compactification. The homological approach based on the Feynman-parametric integral [15] is free from this difficulty.

In this chapter, we are mainly concerned with the locations of the singularities and the behaviors in their neighborhoods, which are independent of the problem of ultraviolet divergence. Hence, for simplicity of descriptions, we hereafter use the unrenormalized Feynman function $f_G(p)$ instead of the renormalized one $\bar{f}_G(p)$, where $f_G(p)$ is *formally* defined to be $f_G(p, 1)$. More precisely,

$$f_G(p) = \frac{(N - 2\mu + r - 1)!}{(r - 1)!} \Lambda^r \int_0^1 d\alpha_0 \int_\Delta d^{N-1}\alpha \times \frac{\alpha_0^{r-1}(1 - \alpha_0)^{N-2\mu-1}}{U^2[\alpha_0 \Lambda + (1 - \alpha_0)\, V]^{N-2\mu+r}} \tag{11-4}$$

where $\Lambda > 0$ and r is a positive integer such that $N - 2\mu + r - 1 \geqq 0$. When rigor is required, we should of course go back to $\bar{f}_G(p)$.

§ 12 LANDAU SINGULARITIES

12-1 Landau Equations

In this section, we investigate the locations of the singularities of the Feynman function $f_G(p)$. Since p is involved only in V, they are determined almost completely by V. The singularities of $f_G(p)$ purely caused by V are called *Landau singularities.* We confine ourselves to Landau singularities until Section 16, in which some exceptional singularities are discussed.

Since the integrand of (11-4) is an analytic function of $\alpha_l\,(l \in |G|)$ except for $\delta\left(1 - \sum_l \alpha_l\right)$, we first carry out one of Feynman-parametric integrations† to eliminate it. Then the integrand is an analytic function of p and $N - 1$ (complex) Feynman parameters together with α_0, and the integration region is compact. As discussed in Appendix A-2, the singularities of $f_G(p)$ can appear when some of Feynman parameters (including α_0) lie at the end points and a pinch occurs for the remaining Feynman parameters. The end-point condition implies that $\alpha_0 = 0$†† and $\alpha_l = 0$ for all $l \in I$ where $I \subset |G|$. Then a pinch can occur if V at $\alpha(I) = 0$ has a double zero under the restriction $\sum_{l \in |G| - I} \alpha_l = 1$. Because of Remark 9-2, V at $\alpha(I) = 0$ equals V_R, where $R \equiv G/I$. By using Lagrange multipler λ, we obtain $V_R = 0$ and $\partial V_R/\partial\alpha_l = \lambda$ for all $l \in |R|$. Since V_R is homogeneous, one finds

$$V_R = \sum_{l \in |R|} \alpha_l\, \partial V_R/\partial\alpha_l = \lambda \tag{12-1}$$

whence $\partial V_R/\partial\alpha_l = 0$ for all $l \in |R|$. Thus:

THEOREM **12-1** [6] A necessary condition for a Landau singularity of $f_G(p)$ is

$$\begin{aligned} \alpha_l &= 0 \quad \text{for all } l \in I \\ \partial V_R/\partial\alpha_l &= 0 \quad \text{for all } l \in |R| \end{aligned} \tag{12-2}$$

where I is a proper subset of $|G|$ and $R \equiv G/I$.

We call the singularity for $R = G$ (i.e., $I = \emptyset$) the *leading singularity* or the *principal singularity*, and those for $R \neq G$ *nonleading singularities.* Since a nonleading singularity of G is the leading singularity of a Feynman

† It is more elegant than doing this to work in a projective space [15].

†† A pinch with respect to α_0 cannot occur because the denominator function is linear in α_0. Furthermore, it is nonvanishing for $\alpha_0 = 1$.

graph R, we may discuss the leading singularity alone without loss of generality. Hence, until the end of Subsection 12-3, we confine ourselves to the leading singularity. Then (12-2) becomes

$$\partial V/\partial\alpha_l = 0 \quad \text{for all } l \in |G| \tag{12-3}$$

In order to investigate (12-3), it is most convenient to make use of the circuit representation (7-26) of V. First, according to (6-19), internal momenta q_l are related to p as

$$\sum_{l\in|G|} [b:l]q_l + p_b = 0 \quad (b \neq a) \tag{12-4}$$

together with $\sum_b p_b = 0$. As remarked in Subsection 6-2, we can arbitrarily choose q_l for all l belonging to a chord set T^* without contradicting (12-4). Hence we may require q_l to satisfy

$$\sum_{l\in|G|} [C_j:l]\,\alpha_l q_l = 0 \quad (j = 1, \ldots, \mu) \tag{12-5}$$

where circuits $C_1, \ldots, C_\mu$ belong to the f-set of circuits corresponding to T^* in the sense of Theorem 2-22, and the Feynman parameters α_l should be understood to assume those specific values corresponding to the pinch. The compatibility between (12-4) and (12-5) is assured by Theorem 3-11 as long as $U \neq 0$.

Because of Theorem 2-19, we can extend (12-5) to

$$\sum_{l\in|G|} [C:l]\,\alpha_l q_l = 0 \quad \text{for all } C \in \boldsymbol{C} \tag{12-6}$$

Then from (7-26), (12-3) reduces to

$$m_l^2 - q_l^2 = 0 \quad \text{for all } l \in |G| \tag{12-7}$$

We call (12-4), (12-5), and (12-7) altogether the *Landau equations*. We may take (7-47) and (12-6) instead of (12-4) and (12-5), respectively, in order to exhibit the duality character of the Landau equations.

THEOREM **12-2** [5] The location of the (leading) singularity of $f_G(p)$ is determined by the Landau equations with $\sum_{l\in|G|} \alpha_l = 1$.

REMARK **12-3** [5] The Landau equations can be derived also from (7-4) together with (7-6) and (7-9). The double zero of Φ occurs when

$$\partial\Phi/\partial\hat{k}_l = 0 \quad \text{for all } l \in T^* \tag{12-8}$$

$$\partial\Phi/\partial\alpha_l = 0 \quad \text{for all } l \in |G| \tag{12-9}$$

On account of (6-18) and (6-24), (12-8) and (12-9) coincide with (12-5) and (12-7), respectively, if we replace k_l by q_l.

REMARK **12-4** [16] If we regard G as an electrical network, α_l as an Ohm resistance, and q_l as an electric current [17], then (12-4) and (12-6) correspond to Kirchhoff's first law and second law, respectively. Furthermore,

$$Q \equiv \sum_{l \in |G|} \alpha_l m_l^2 - V \tag{12-10}$$

corresponds to the power dissipated in G.

According to Theorem 3-11, if $U \neq 0$ then the simultaneous equations (12-4) and (12-5) can be solved uniquely with respect to q_l. Indeed, the unique solution is given by

$$q_l = Y_l \quad \text{for all } l \in |G| \tag{12-11}$$

because of (9-20) together with (12-6) [18]. From (9-25), therefore, we have

$$q_l = U^{-1} \sum_{b \in g-a} W_l^{(a|b)} p_b \tag{12-12}$$

We call any solution of the Landau equations such that $U = 0$ but $\prod_{l \in |G|} \alpha_l \neq 0$, a *peculiar solution*. In order for a peculiar solution to exist, (12-12) has to be an indefinite form 0/0. Hence if p_b $(b \neq a)$ are indepedent we have

$$W_l^{(a|b)} = 0 \quad \text{for all } l \in |G| \text{ and all } b \in g - a \tag{12-13}$$

In this case, q_l can be linearly independent of external momenta, that is, the condition for the singularity generally admits extra freedoms for q_l. The pinch in such a case becomes a nonsimple pinch in the momentum-space Feynman integral [19].

Example According to (9-27), $Y_1, \ldots, Y_5$ of the self-energy graph shown in Fig. 12-1 are as follows:

$$UY_1 = (\alpha_2\alpha_4 + \alpha_4\alpha_5 + \alpha_5\alpha_2 + \alpha_2\alpha_3)p$$

$$UY_2 = (\alpha_1\alpha_3 + \alpha_3\alpha_5 + \alpha_5\alpha_1 + \alpha_1\alpha_4)p$$

$$UY_3 = (\alpha_2\alpha_4 + \alpha_4\alpha_5 + \alpha_5\alpha_2 + \alpha_1\alpha_4)p$$

$$UY_4 = (\alpha_1\alpha_3 + \alpha_3\alpha_5 + \alpha_5\alpha_1 + \alpha_2\alpha_3)p$$

$$UY_5 = (\alpha_2\alpha_3 - \alpha_1\alpha_4)p$$

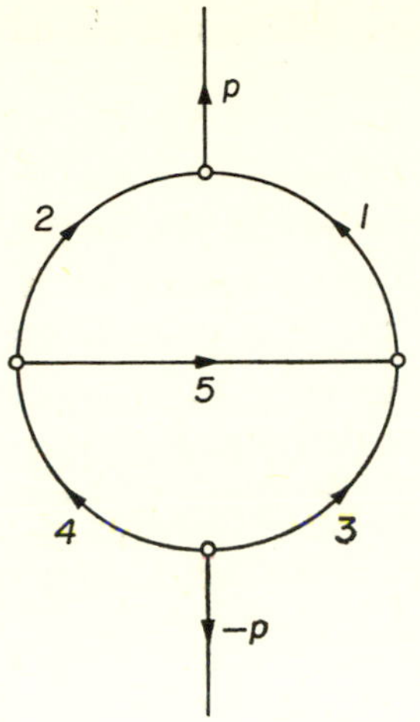

Fig. 12-1 A Feynman graph yielding a peculiar solution to the Landau equations.

with

$$U = (\alpha_1 + \alpha_2)(\alpha_3 + \alpha_4) + (\alpha_3 + \alpha_4)\alpha_5 + \alpha_5(\alpha_1 + \alpha_2)$$

The condition $U = 0$ for $\prod_l \alpha_l \neq 0$ is compatible with the pinch condition if

$$\alpha_1\alpha_3 + \alpha_3\alpha_5 + \alpha_5\alpha_1 = \alpha_2\alpha_4 + \alpha_4\alpha_5 + \alpha_5\alpha_2 = -\alpha_1\alpha_4 = -\alpha_2\alpha_3$$

In what follows, we consider the ordinary case $U \neq 0$ only. Let d be the number of linearly independent external momenta p_b when they are fixed to general values, that is,

$$d = \min(n - 1, 4) \tag{12-14}$$

because the momentum space is of four dimensions.

Since each internal momentum q_l is a linear combination of external momenta, the numbers of independent Landau equations (12-4), (12-5), and (12-7) are $d(M - 1)$, $d\mu$, and N, respectively. Hence the total number is $(d + 1)N$. On the other hand, the numbers of unknown quantities q_l and α_l $\left(\text{with } \sum_l \alpha_l = 1\right)$ are dN and $N - 1$, respectively. Hence the total number is $(d + 1)N - 1$. The deficiency of one unknown quantity leads to one relation for external momenta. In the Landau equations, however, the Feynman parameters are involved only in (12-5). Accordingly, $d(N - \mu)$

$+ N$ equations (12-4) and (12-7) involve dN unknown quantities q_l. Therefore unless

$$N - d\mu \leqq 1 \tag{12-15}$$

there have to exist two or more relations for external momenta. As remarked in Appendix A-2, however, the singularity of an analytic function cannot be such a low-dimensional manifold; it should be a submanifold of a nonleading singularity. Thus we conclude that if (12-15) is not satisfied, the leading singularity is absent. The same condition has to be satisfied in every sub-Feynman-graph of G, because every subsystem of the Landau equations also should not impose excessive relations on momenta.

THEOREM **12-5** A necessary condition for the presence of the leading singularity is that the inequality

$$N(H) \leqq d(H)\mu(H) + 1 \tag{12-16}$$

holds for every sub-Feynman-graph H of G.

For example, no leading singularity exists for single-loop (i.e., $\mu = 1$) Feynman graphs with $N > \max(n, 5)$. It may be expected that if the condition of Theorem 12-5 is satisfied then the leading singularity will be present on some Riemann sheets except for some possible exceptional values of m_l $(l \in |G|)$. If $N = d\mu + 1$ then the internal momenta q_l are determined without solving (12-5), which determines α_l $(l \in |G|)$ alone. For example, this is the case for single-loop Feynman graphs with $2 \leqq N = n \leqq 5$.

12-2 Dual-Graph Analysis

It is in general very difficult to solve the Landau equations analytically. It is therefore helpful to analyze them in a geometrical way.

As defined in Section 4, the Feynman graph G is called planar if G' is planar, where G' is obtained from G by regarding all the free ends of the external lines of G as a new common vertex "infinity". In this case, G has a dual graph $\tilde{G}$. By definition, $\tilde{G}$ has $\mu + n$ vertices.†

We identify each line $\tilde{l}$ of $\tilde{G}$ with an "arrow" representing the related complex 4-momentum (q_l if $\tilde{l}$ corresponds to an internal line l of G, or p_b if it corresponds to an external line incident with b), that is, $\tilde{l}$ is regarded as a complex d-dimensional geometrical vector. Then the Landau equations have the following geometrical meanings.

† Without loss of generality, we suppose that every external vertex is an end point of only one external line.

The conservation law (12-4) expresses that the momenta incident with each vertex in G form a closed polygon in the complex d-dimensional space embedding $\tilde{G}$. Certain μ equations of (12-6) imply that all vectors incident with each vertex $\tilde{a}$ of $\tilde{G}$ are linearly dependent if $\tilde{a}$ corresponds to a circuit in G. Finally, (12-7) assigns a "length" m_l in the Minkowski-metric sense to each vector q_l.

A more concrete analysis of using dual graphs is presented in Subsection 13-2.

As remarked in Section 4, dual graphs are not necessarily unique. In that case, according to Theorem 4-10 and Definition 4-8, a dual graph $\tilde{G}$

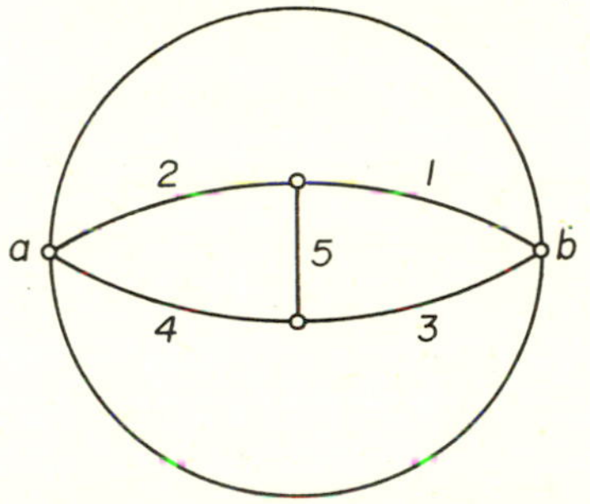

Fig. 12-2 A dual graph to Fig. 12-1.

consists of two connected subgraphs $\tilde{H}$ and $\tilde{H}^*$ such that there are only two vertices $\tilde{a}$ and $\tilde{b}$ belonging to both $\tilde{H}$ and $\tilde{H}^*$, provided that G' is non-separable. It is possible, therefore, to rotate either $\tilde{H}$ or $\tilde{H}^*$ in the geometrical sense without violating the Landau equations. This freedom is closely related to the existence of a peculiar solution, because then internal momenta may not lie in the space spanned by external momenta. For example, a dual graph of Fig. 12-1 is Fig. 12-2; the central part can be rotated.

12-3 Behavior near the Singularity

In this subsection, we investigate the behavior of $f_G(p)$ in a neighborhood of a singularity. Let ΔV be a variation of V when p is varied by Δp from the position of the singularity while the Feynman parameters are held fixed. Furthermore, we denote the value of α_l at the position of singularity by

$\bar{\alpha}_l$ and the difference $\alpha_l - \bar{\alpha}_l$ by η_l; of course $\sum_{l\in|G|} \eta_l = 0$. When ΔV and η_l ($l \in |G|$) are infinitesimal, we can write

$$V \cong \Delta V + \tfrac{1}{2} \sum_{l,l'\in|G|} a_{ll'}\eta_l\eta_{l'} \tag{12-17}$$

with

$$a_{ll'} \equiv \partial^2 V/\partial\alpha_l\partial\alpha_{l'}|_{\alpha=\bar{\alpha}} \tag{12-18}$$

in a neighborhood of the singularity, where $\alpha = \bar{\alpha}$ means $\alpha_l = \bar{\alpha}_l$ for all $l \in |G|$. Since the Landau equations are invariant under the scale change of Feynman parameters, the second term of (12-17) vanishes when $\eta_l = \lambda\bar{\alpha}_l$ ($\lambda \neq 0$), whence

$$\det(a_{ll'}\,|l, l' \in |G|) = 0 \tag{12-19}$$

This variation, however, is not admissible because then $\sum_l \eta_l = \lambda \neq 0$. If there is no special reason, therefore, the rank of the matrix $(a_{ll'})$ should be $N - 1$, the dimension of the admissible variation. By diagonalizing the matrix $(a_{ll'})$, the second term of (12-17) can be written as

$$\tfrac{1}{2} \sum_{j=1}^{N-1} b_j\zeta_j^2, \quad \Big(\prod_j b_j \neq 0\Big) \tag{12-20}$$

Then the integral (11-4) may be approximated by

$$\int_0^1 d\alpha_0\alpha_0^{r-1} \prod_{j=1}^{N-1} \left(\int_{-1}^{1} d\zeta_j\right) \Big(\Delta V + \tfrac{1}{2}\sum_j b_j\zeta_j^2 + \alpha_0 \Lambda\Big)^{-N+2\mu-r} \tag{12-21}$$

near the singularity (apart from a multiplicative constant). On carrying out the integrations successively by taking account of the contribution from a neighborhood of $\zeta_1 = \ldots = \zeta_{N-1} = \alpha_0 = 0$ only, we finally find that $f_G(p)$ behaves like

$$\begin{array}{ll} (\Delta V)^k \log(\Delta V) & \text{for } k = 0, 1, 2, \ldots \\ (\Delta V)^k & \text{otherwise} \end{array} \tag{12-22}$$

where

$$k \equiv (-N + 2\mu - r) + \tfrac{1}{2}(N - 1) + r = \tfrac{1}{2}(4\mu - N - 1) \tag{12-23}$$

Since $N \leqq 4\mu + 1$ because of (12-15) together with (12-14), we find $k \geqq -1$, that is, the strongest possible singularity is a simple pole, and multiple poles are absent except for some special values of internal masses. For $k \geqq -\tfrac{1}{2}$, the singularity is a branch point.

The above discussion has been made for the leading singularity. If we extend it to the case of a nonleading singularity, some care has to be taken. We have to take account of the $N(I)$ integrations over α_l $(l \in I)$ and the multiplicity $2\mu(I)$ of the zero of U^2 [see (9-1)].

Since V varies with α_l $(l \in I)$ approximately linearly, (12-23) should be replaced by

$$\begin{aligned} k &= -N + 2\mu + \tfrac{1}{2}[N(R) - 1] + N(I) - 2\mu(I) \\ &= \tfrac{1}{2}[4\mu(R) - N(R) - 1] \end{aligned} \tag{12-24}$$

For the case in which nonzero-spin particles are present, (12-24) presents a lower bound on k.

§ 13 REAL SINGULARITIES

13-1 General Consideration on Real Singularities

It is extremely difficult to discuss the detailed analytic properties of the general Feynman function. Since the quantity of physical interest is not the Feynman function but the Feynman integral, we investigate the singularities of $f_G(p)$ with $\varepsilon \to +0$ in this section.

The singularities of $f_G(p)$ are those of $f_G(p)$ which are characterized by the reality of V. More precisely, we consider the Landau singularities of $f_G(p)$ specified by Theorem 12-1 under additional conditions

$$\alpha_l > 0 \quad \text{for all } l \in |R| \tag{13-1}$$

$$\operatorname{Im} s_h = 0 \quad \text{for all } h \subset g \tag{13-2}$$

where s_h is defined by (11-1). If they lie on the physical sheet, then they are called *real singularities*. The singularities of $f_G(p)$ are nothing but real singularities of $f_G(p)$.

Let D_R be the region defined by (13-2), and $D_0 \equiv D_V \cap D_R$, where D_V is defined by (11-2). We may also characterize D_0 as the set of points such that $V > 0$ for all $\alpha \in \Delta$ because V is continuous in Δ and $V > 0$ in some subregion of Δ (see Remark 9-2). Evidently, no real singularities exist in D_0. We note that D_0 is not necessarily non-empty.

Real singularities are much simpler than the general Landau singularities. For example, any peculiar solution to the Landau equations cannot correspond to a real singularity because (13-1) implies $U_R > 0$.

THEOREM **13-1** [20] If all of the conditions (12-2), (13-1), and (13-2) are satisfied, then a real singularity is always present (except for some possible exceptional values of internal masses).

Proof As in (12-17) and (12-20), we may write

$$V_R \cong \Delta V_R + \tfrac{1}{2} \sum_{j=1}^{N(R)-1} b_j \zeta_j^2 \tag{13-3}$$

near the point which satisfies (12-2). Since all Feynman parameters and invariant squares are real, ΔV_R, b_j, and $\zeta_j (j = 1, \ldots, N(R) - 1)$ are all real. Except for the case $\prod_j b_j = 0$, as $\varepsilon \to +0$ the solutions to equation $V_R - i\varepsilon = 0$ approach to the real axis of ζ_j from both sides. Thus a pinch occurs at every stage of integrations. Therefore a real singularity is present. In the exceptional case $\prod_j b_j = 0$, the singularity will also be present except for particular values of internal masses because the singularity manifold is a closed set of points. q.e.d.

Hereafter in this section, we regard $f_G(p)$ as a distribution of only one particular invariant square s, and we denote it by $f_G(s)$. Here we choose as s one of invariant squares such that $\partial V/\partial s < 0$ in a neighborhood of the real singularity under consideration. [This may be always possible as seen from the modified cut-set representation (7-46) of V (see Section 17 for more detail).] If $\partial V/\partial s < 0$ holds for all $\alpha \in \Delta$, then $f_G(s)$ can be regarded as a boundary value of an analytic function $f_G(s)$ from the upper half plane of s. This function is holomorphic in $\operatorname{Im} s \neq 0$.

Now, we consider a real singularity specified by (12-2), (13-1), and (13-2). We denote the solution of (12-2) by $s = \bar{s}$ and $\alpha = \bar{\alpha}$ (i.e., $\alpha_l = \bar{\alpha}_l$ for all $l \in |R|$). We write

$$V_R(s, \alpha) = A(\alpha)s + B(\alpha) \tag{13-4}$$

Then by assumption $A(\alpha)$ is negative in a neighborhood of $\alpha = \bar{\alpha}$. Putting

$$C(\alpha) \equiv A(\alpha)\bar{s} + B(\alpha) \tag{13-5}$$

we have

$$V_R(s, \alpha) = A(\alpha)(s - \bar{s}) + C(\alpha) \tag{13-6}$$

with

$$\partial C/\partial \alpha_l|_{\alpha=\bar{\alpha}} = 0 \quad (l \in |R|) \tag{13-7}$$

The following theorem assures that V_R has no maximal zero point, though it may have maximal nonzero points.

THEOREM **13-2** [16] As a function of $\alpha \in \Delta$, V has no maximal zero point.

Proof Let $s = \bar{s}$ and $\alpha_l = \bar{\alpha}_l \, (l \in |G|)$ be the solution of (12-3), and define $\eta_l \equiv \alpha_l - \bar{\alpha}_l$ with $\sum_l \eta_l = 0$ as before. If $\alpha = \bar{\alpha}$ is a maximal point of $V(\bar{s}, \alpha)$, then we always have

$$V(\bar{s}, \bar{\alpha} + \eta) \leqq 0 \tag{13-8}$$

for $|\eta| \equiv \sum_l |\eta_l|$ infinitesimal. On substituting the Landau equations (12-6) and (12-7) in the circuit representation (7-26) of V, we find

$$V(\bar{s}, \bar{\alpha} + \eta) = [U(\bar{\alpha})]^{-1} \sum_{C \in C} U_C(\bar{\alpha}) \Big(\sum_{l \in |G|} [C:l]\eta_l \bar{q}_l \Big)^2 + O(|\eta|^3) \tag{13-9}$$

where $q = \bar{q}$ is the solution of the Landau equations. Let

$$\eta_l = \eta_0(\delta_{lk} - \bar{\alpha}_l) \quad (l \in |G|) \tag{13-10}$$

Then we obtain

$$V(\bar{s}, \bar{\alpha} + \eta) = \frac{\eta_0^2}{U} \left. \frac{\partial U}{\partial \alpha_k} \right|_{\alpha = \bar{\alpha}} \bar{q}_k^2 + O(|\eta|^3) \tag{13-11}$$

with the aid of (12-6) and (7-33). Since $\bar{q}_k^2 = m_k^2 > 0$ because of (12-7), (13-11) is positive in contradiction with (13-8). q.e.d.

From this theorem, we see that $\alpha = \bar{\alpha}$ is a minimal or a saddle point of $V_R(\bar{s}, \alpha)$. If $\alpha = \bar{\alpha}$ is a minimal point of $V_R(\bar{s}, \alpha)$, then we call $s = \bar{s}$ a *threshold.* If $s = \bar{s}$ is a threshold, we have

$$C(\alpha) \geqq 0 \quad \text{near } \alpha = \bar{\alpha} \tag{13-12}$$

together with $A(\alpha) < 0$. Therefore, in a neighborhood of $s = \bar{s}$, we see from (13-6) that when $s < \bar{s}$

$$V_R(s, \alpha) \geqq 0 \quad \text{for } any\ \alpha \text{ near } \alpha = \bar{\alpha} \tag{13-13}$$

but when $s > \bar{s}$

$$V_R(s, \alpha) < 0 \quad \text{for } some\ \alpha \text{ near } \alpha = \bar{\alpha} \tag{13-14}$$

Thus, according to (7-66), the contribution to $\operatorname{Im} f_G(s)$ from a neighborhood of $\alpha = \bar{\alpha}$ is zero for $s < \bar{s}$ but in general non-zero for $s > \bar{s}$, that is, a new contribution to the imaginary part begins to occur at $s = \bar{s}$. Physically, this fact can be interpreted as the appearance of a new competing process if we are in the physical region [see (7-65)].

As shown in Subsection 12-3, singularities of $f_G(p)$ are poles and branch points; it has no essential singularities. If we differentiate $f_G(p)$ with respect to a relevant variable s sufficiently many times, then it becomes infinite at the position of the singularity. Therefore we can characterize non-singular points by the property that all derivatives with respect to s are finite. In this way, we can define the singularities of $\operatorname{Re} f_G(s)$ and $\operatorname{Im} f_G(s)$.

THEOREM **13-3** [16] A necessary and sufficient condition for $\operatorname{Re} f_G(s)$ being singular at $s = \bar{s}$ is that $s = \bar{s}$ is a threshold.

Proof If $\alpha = \bar{\alpha}$ is a saddle point of $V_R(\bar{s}, \alpha)$, then in (13-3) there exist j and j' such that $b_j > 0$ and $b_{j'} < 0$, provided that $\prod_i b_i \neq 0$. Then V_R is linear with respect to new variables $u = \zeta_j + \zeta_{j'}$ and $v = \zeta_j - \zeta_{j'}$, and the Jacobian is a constant. According to (A-16) of Appendix A, the integrand of Re $f_G(s)$ is expressed in terms of Hadamard's finite part. Since neither $u = 0$ nor $v = 0$ is an end point of the integration region, therefore, the integrations over u and v are well defined. The same is true also for any finite-order derivative of Re $f_G(s)$. Thus Re $f_{G(s)}$ is not singular at $s = \bar{s}$.

For the case of a threshold, all of b_j ($j = 1, \ldots, N(R) - 1$) are positive, provided that $\prod_i b_i \neq 0$. Then with a new variable $r = \sum_j b_j \zeta_j^2$, the integration range is $r \geqq 0$, whence $(d/ds)^k$ Re $f_G(s)$ becomes divergent if k is sufficiently large, because we no longer have Hadamard's finite part. Thus Re $f_G(s)$ is singular at $s = \bar{s}$.

We may extend the above reasoning to the exceptional case $\prod_i b_i = 0$ by considering higher order terms. q.e.d.

THEOREM **13-4** At $s = \bar{s}$, Im $f_G(s)$ is always singular.

Proof From the discussion of the threshold behavior, it is clear that Im $f_G(s)$ is singular at $s = \bar{s}$ if it is a threshold. For the non-threshold case, Theorems 13-1 and 13-3 imply that Im $f_G(s)$ has to be singular at $s = \bar{s}$. q.e.d.

Example Consider an integral

$$f(s) \equiv \int_{-1}^{1} d\alpha_1 \int_{-1}^{1} d\alpha_2 \, (\alpha_1 \alpha_2 - s - i\varepsilon)^{-2} \quad (\varepsilon \to +0)$$

It has a saddle point at $s = 0$. By direct calculation, we find

$$\mathrm{Re} f(s) = \frac{2}{s} \log \left| \frac{s+1}{s-1} \right| = 4 \left(1 + \frac{s^2}{3} + \frac{s^4}{5} + \ldots \right)$$

$$\mathrm{Im} f(s) = 2\pi s^{-1} [\theta(s-1) - \theta(s+1)]$$

Thus at $s = 0$ Re $f(s)$ is non-singular but Im $f(s)$ is singular.

13-2 Normal and anomalous thresholds

In this subsection, we investigate the properties of thresholds in more detail.

DEFINITION **13-5** A set of complex Minkowski vectors $\{p_1, \ldots, p_k\}$ is called *euclidean* if

$$\left(\sum_{j=1}^{k} c_j p_j \right)^2 \geqq 0 \tag{13-15}$$

for any real numbers $c_1, \ldots, c_k$.

THEOREM **13-6** If Minkowski vectors $p_1, \ldots, p_k$ are all real but non-zero and if they form a euclidean set, then they are timelike or lightlike (i.e., $p_i^2 \geqq 0$) and parallel to each other.

Proof The first statement directly follows from (13-15) by setting $c_j = \delta_{ji}$. We consider two arbitrary vectors, say p_1 and p_2. Since they are real and not spacelike, we have $p_1^{(0)} \neq 0$ and $p_2^{(0)} \neq 0$. Hence we can find a linear combination $c_1 p_1 + c_2 p_2 (c_1 \neq 0, c_2 \neq 0)$ such that $c_1 p_1^{(0)} + c_2 p_2^{(0)} = 0$. Then (13-15) and the reality require $c_1 p_1 + c_2 p_2 = 0$, that is, p_1 and p_2 are parallel. q.e.d.

We denote the set of external vertices of the reduced graph R by g^R, the external momenta outgoing from each external vertex $b \in g^R$ by p_b^R, and the number of linearly independent p_b^R by $d(R)$. From Theorem 13-6, we see that $\{p_b^R \mid b \in g^R\}$ can be euclidean in the physical region (i.e., p_b^R real) only when $d(R) = 1$.

THEOREM **13-7** [18] Let $p_b^R = \bar{p}_b^R \, (b \in g^R)$ be the solution of the Landau equations and $s = \bar{s}$ be the corresponding one. If $\{\bar{p}_b^R \mid b \in g^R\}$ is euclidean, $s = \bar{s}$ is a threshold.

Proof Since all internal momenta $\bar{q}_l$ in R are linear combinations of $\bar{p}_b^R$, they form a euclidean set. Then (13-9) tells us that $V_R(\bar{s}, \alpha) \geqq 0$ in a neighborhood of $\alpha = \bar{\alpha}$. q.e.d.

The inverse proposition of Theorem 13-7 is true only for $\mu = 1$. Thresholds which do not correspond to euclidean sets can easily be found for $\mu \geqq 2$.

If R is a planar Feynman graph and if $\{\bar{p}_b^R \mid b \in g^R\}$ is euclidean, then it is very convenient to employ the dual-graph analysis mentioned in Subsection 12-2. We can now embed the dual graph $\tilde{R}$ in a real $d(R)$-dimensional euclidean space. We can give the Landau equations more precise geometrical meanings: (12-4) means that the momenta incident with each vertex of R form a closed euclidean polygon (not planar in general); (12-7) implies that the *euclidean length* of the vector $\bar{q}_l$ is equal to $m_l > 0$; (12-6) leads to the property that each vertex $\tilde{a}$ of $\tilde{R}$ corresponding to a circuit in R lies in the subspace which is determined by its adjacent vertices, and moreover (13-1) means that $\tilde{a}$ lies *inside* the polyhedron whose vertices are its adjacent ones†.

† For example, for $N(R) = 3$ the Feynman parameters can be regarded as the triangular coordinates in the subspace.

Example [21] The dual graph of the "triangle graph" shown in Fig. 13-1 is Fig. 13-2. A dual graph of the "square graph" ($\mu = 1$, $N = n = 4$) is a pentahedron.

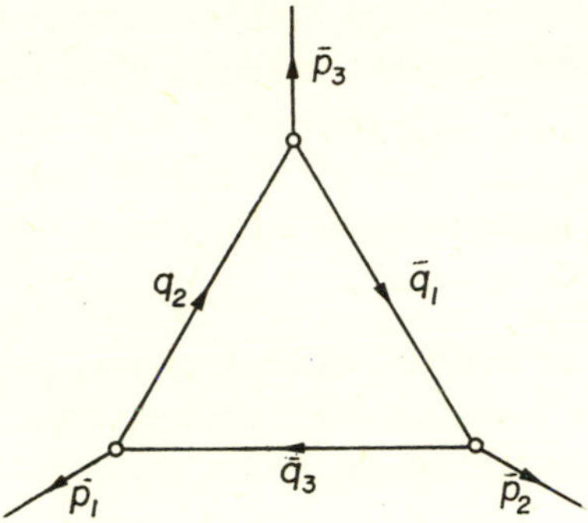

Fig. 13-1 The triangle graph.

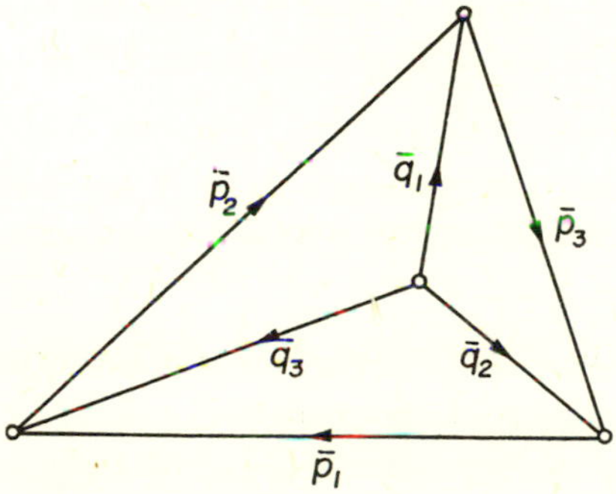

Fig. 13-2 A dual graph to Fig. 13-1.

The simplest case is $d(R) = 1$. In this case, from $\bar{q}_l^2 = m_l^2 > 0$ and $\bar{q}_l$ being parallel to $\bar{p}^R$, we have $(\bar{p}^R)^2 > 0$, whence $s = \bar{s}$ is a threshold because of Theorem 13-7. We call a threshold of $d(R) = 1$ a *normal threshold*, and that of $d(R) \geqq 2$ an *anomalous threshold*. For a normal threshold, we know $N(R) \leqq \mu(R) + 1$ form (12-15), and therefore the number of the vertices of R is $N(R) - \mu(R) + 1 \leqq 2$. Since $m_l > 0$ for all $l \in |R|$, the existence of loop lines is excluded by (12-7). Hence the only possible graph R consists of $N(R)$ parallel lines alone which link the two external vertices of R, that is, $|R|$ is a cut-set S in G. An example of R is shown in Fig. 13-3. Let a_S and b_S be the two vertices of R and define $[S : l]$ by $[a_S : l]$ of R (see

Definition 2-12). We denote the external momentum outgoing from b_S in R by $\bar{p}^R$. Then the Landau equations read

$$\begin{aligned}
&\sum_{l\in S} [S:l]\,\bar{q}_l = \bar{p}^R \\
&[S:l]\,\bar{\alpha}_l\bar{q}_l = [S:l']\,\bar{\alpha}_{l'}\,\bar{q}_{l'} \qquad (l, l' \in S) \\
&\bar{q}_l^2 = m_l^2 \quad (l \in S)
\end{aligned} \tag{13-16}$$

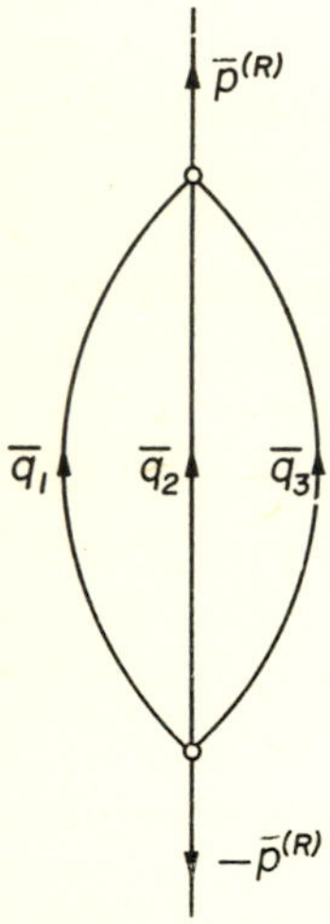

Fig. 13-3 A reduced graph corresponding to a normal threshold.

It is straightforward to solve (13-16) under the condition (13-1). We find

$$\begin{aligned}
&\bar{\alpha}_l = m_l^{-1} / \sum_{k\in S} m_k^{-1} \quad (l \in S) \\
&\bar{q}_l = [S:l]\left(m_l / \sum_{k\in S} m_k\right)\bar{p}^R \quad (l \in S) \\
&\bar{s} \equiv (\bar{p}^R)^2 = \left(\sum_{l\in S} m_l\right)^2
\end{aligned} \tag{13-17}$$

together with

$$A(\bar{\alpha}) = -\left(\sum_{l\in S} m_l^{-1} \sum_{k\in S} m_k\right)^{-1} < 0 \tag{13-18}$$

Since every cut-set S in G yields a reduced graph $R \equiv G/(|G| - S)$, we have the following theorem (see Subsection 12-3).

THEOREM **13-8** For every intermediate state S in a channel $(h \mid g - h)$ [i.e., $S \in \boldsymbol{S}(h \mid g - h)$], $f_G(s)$ has a normal threshold at

$$s = \bar{s} = \left(\sum_{l\in S} m_l\right)^2 \tag{13-19}$$

if we define s by $\left(\sum_{b\in h} p_b\right)^2$. Near $s = \bar{s}$, $f_G(s)$ behaves like $(s - \bar{s})^{-1}$ for $N(S) = 1$, $(s - \bar{s})^k$ for $N(S) = 2, 4, \ldots$, and $(s - \bar{s})^k \log(s - \bar{s})$ for $N(S) = 3, 5, \ldots$, where $k = [3N(S) - 5]/2$.

For an anomalous threshold, $|R|$ has to include an intermediate state as a *proper* subset, but every set having such a property does not necessarily correspond to an anomalous threshold. In order to investigate in what case an anomalous threshold appears, we should consider variations of the remaining Feynman parameters α_l ($l \in I \equiv |G| - |R|$). We consider the disconnected sub-Feynman-graph $H \equiv G - |R|$; of course $I = |H|$.

THEOREM **13-9** [18] When all momenta and Feynman parameters relating to R assume the values satisfying the Landau equations, we have

$$V \cong V_H \tag{13-20}$$

to first order in α_l ($l \in |H|$), where V_H stands for the V function of H with external momenta

$$p_b^H \equiv p_b + \sum_{l\in|R|} [b:l]\,\bar{q}_l \quad \text{for } b \in v(H) \tag{13-21}$$

Proof In H, the momentum conservation at each vertex b is written as

$$\sum_{l\in|H|} [b:l]\, q_l + p_b^H = 0 \tag{13-22}$$

We consider the circuit representation (7-26) of V. The first term of V reduces to that of V_H when $q_l = \bar{q}_l$ for all $l \in |R|$. We investigate the behavior of

$$(U_C/U)\left(\sum_{l\in|H|} [C:l]\,\alpha_l q_l\right)^2 \tag{13-23}$$

at $\alpha_k = \bar{\alpha}_k$ ($k \in |R|$) for every circuit C in G. If $C \not\subset |H|$ then both U and U_C are of $\mu(H)$-th order with respect to α_l ($l \in |H|$) because of (9-1), whence (13-23) is of at least second order. If $C \subset |H|$ then U is of $\mu(H)$-th order but U_C is of $[\mu(H) - 1]$-th order. Furthermore, (9-1) yields $U_C/U \cong U_{H/C}/U_H$. Thus the second term of V becomes that of V_H to first order of α_l($l \in |H|$). q.e.d.

Let $\bar{s}$ be the minimum of $\left(\sum_{l\in S} m_l\right)^2$ when S runs over all intermediate states in a channel $(h \mid g - h)$ such that $N(S) \geqq 2$ (we exclude S consisting of a cut-line because it is trivial to take it into account). We call $\bar{s}$ the *lowest normal threshold* in the channel $(h \mid g - h)$.

THEOREM **13-10** [16] For a strongly connected graph G, let $\bar{s}$ be the lowest normal threshold in a channel $(h \mid g - h)$ and S be the corresponding intermediate state. An anomalous threshold $s = \hat{s}$ such that $s_0 \leqq \hat{s} < \bar{s}$ is present if there exists s_0 such that

$$V(s_0, \alpha) \geqq 0 \quad \text{for all } \alpha \in \Delta \tag{13-24}$$

and if

$$V_H(\bar{s}, \alpha) < 0 \quad \text{for some } \alpha \in \Delta_H \tag{13-25}$$

where Δ_H is the Δ with $\alpha(S) = 0$, and V_H is the V function of $H \equiv G - S$ with external momenta (13-21) $(|R| \equiv S)$.

Proof Let

$$\hat{s} \equiv \sup \quad \{s \mid V(s, \alpha) \geqq 0 \quad \text{for all } \alpha \in \Delta\} \tag{13-26}$$

First, we observe that there exists α such that $V(\hat{s}, \alpha) = 0$, because otherwise the continuity of V implies the existence of $\hat{s}' > \hat{s}$ such that $V(\hat{s}', \alpha) \geqq 0$ for all $\alpha \in \Delta$ in contradiction with (13-26). Therefore $s = \hat{s}$ presents a minimal zero point of V, that is, it is a threshold. Furthermore, we observe that $\hat{s} \geqq s_0$ on account of (13-24), that $\hat{s} \leqq \bar{s}$ because $\partial V/\partial \bar{s} < 0$ at the value of α giving the normal threshold (see (13-18)), and that $\hat{s} \neq \bar{s}$ because of Theorem 13-9 and (13-25). Thus there is a threshold $s = \hat{s}$ such that $s_0 \leqq \hat{s} < \bar{s}$, and it cannot be a normal threshold. q.e.d.

Example Let G be the triangle graph shown in Fig. 13-4. It is straightforward to show that

$$V(s, \alpha) \geqq 0 \quad \text{for } s \equiv (p_a + p_b)^2 < (m_1 - m_2)^2$$

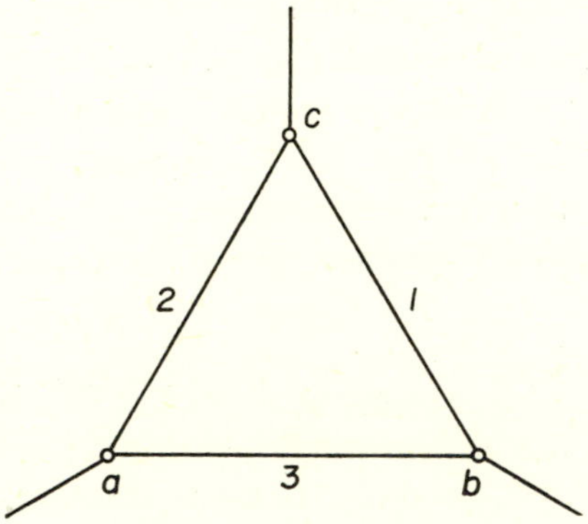

Fig. 13-4 The triangle graph.

if $p_a^2 \leqq (m_2 + m_3)^2$ and $p_b^2 \leqq (m_3 + m_1)^2$ (stability conditions). Therefore, a sufficient condition for the existence of an anomalous threshold $s = \hat{s}$ in $(m_1 - m_2)^2 \leqq s < (m_1 + m_2)^2$ is

$$V_H = \alpha_3(m_3^2 - \bar{q}_3^2) < 0$$

where

$$\bar{q}_3 = \bar{p}_a - \frac{m_2}{m_1 + m_2}(\bar{p}_a + \bar{p}_b)$$

with $(\bar{p}_a + \bar{p}_b)^2 = (m_1 + m_2)^2$. That is to say, we have

$$m_1(\bar{p}_a^2 - m_2^2 - m_3^2) + m_2(\bar{p}_b^2 - m_3^2 - m_1^2) > 0 \qquad (13\text{-}27)$$

A similar consideration also applies to the square graph if s is a square of a sum of two external momenta associated with adjacent vertices.

As described in Appendix B, (13-27) is indeed a necessary and *sufficient* condition for the existence of the anomalous threshold in s in the triangle and square graphs [2, 21]. Furthermore, we note that (13-27) is equivalent to the condition under which the dispersion relation for forward scattering (see next chapter) cannot be proven in the axiomatic field theory, but we omit the detail [16].

§ 14 PHYSICAL-REGION SINGULARITIES

This section is only of physical interest. We use some physical terms without explaining them explicitly.

14-1 Multiple Scattering

Physical-region singularities are real singularities for which all external momenta are real [of course this condition is stronger than (13-2)]. Normal thresholds are physical-region singularities, but anomalous thresholds are not in general. As seen later, if all external momenta satisfy the "stability condition" (i.e., p_a^2 is less than the lowest normal threshold M_a^2 in the channel $(a \mid g - a)$), then there exist no physical-region singularities other than normal thresholds. Hence we should have $p_a^2 > M_a^2$ for at least one vertex a. If no unstable particles are involved, that is possible on the mass shell only when two or more external lines are incident with a. Thus, in this section, it is convenient to distinguish various external lines incident with the same vertex from each other.

As in Section 12, we consider the leading singularity alone. The leading physical-region singularity appears if and only if the Feynman graph G is consistently interpreted as a real multiple scattering. The simplest non-trivial example is shown in Fig. 14-1.

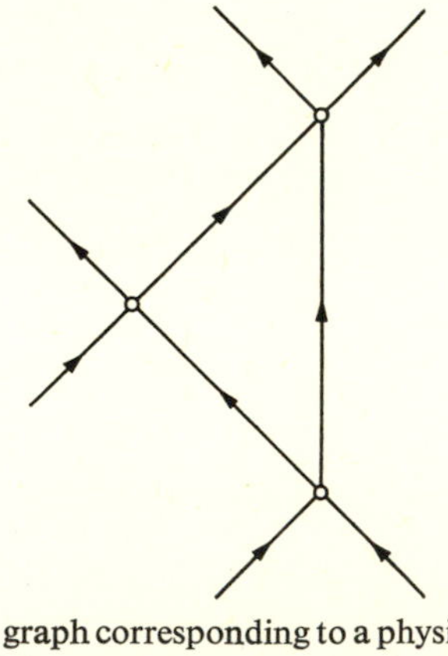

Fig. 14-1 A Feynman graph corresponding to a physical-region singularity. The 0-th component of the momentum associated with each line is positive along its orientation.

THEOREM **14-1** [20] The leading physical-region singularity exists if and only if G can be consistently interpreted as the trajectories of classical, relativistic particles in the position space when we identify the Feynman parameters α_l with $\tau_l/m_l > 0\,(l \in |G|)$, where τ_l stands for the proper time interval of the particle corresponding to l.

Proof Given the Landau equations, we can choose the orientation of each line l in such a way that $q_l^{(0)} \geqq 0$. For each vertex b, we associate a 4-vector x_b with it. We can put

$$\alpha_l q_l = -\sum_{b \in |G|} [b : l]\, x_b \tag{14-1}$$

because then the circuit equations (12-5) are identically satisfied on account of (2-8). On substituting τ_l/m_l for α_l in (14-1), we find that x_b can be interpreted as the space-time position of the "event" b. The other Landau equations (12-4) and (12-7) are the 4-momentum conservation and the mass-shell conditions, respectively. Thus the Feynman graph is regarded as trajectories of classical, relativistic particles. Furthermore, in this case, the singularity is always present because of Theorem 13-1. q.e.d.

It is important to note that the time ordering is introduced between any two adjacent vertices because $\tau_l > 0$. Thus the set $\{x_b \mid b \in v(G)\}$ has a semiordering. Hence there are at least one first vertex and at least one last one. For such a vertex a, we have to have $p_a^2 > M_a^2$ to satisfy the 4-momentum conservation law, as mentioned at the beginning of this subsection.

The above theorem can be made more precise if we consider the singularity directly in the position space [22]. First, we rewrite the Landau equations in terms of the relative positions $x_b^a \equiv x_b - x_a\,(b \in v(G) - a)$ and the inverse Feynman parameters $\beta_l\,(l \in |G|)$. According to (14-1), (12-4) becomes

$$\sum_{c \neq a} w_{bc} x_c^a - p_b = 0 \quad (b \neq a) \tag{14-2}$$

where

$$w_{bc} \equiv \sum_l \,[b : l]\,[c : l]\,\beta_l \tag{14-3}$$

The mass-shell equations (12-7) are rewritten as

$$\beta_l^2 \Big(\sum_{b \neq a} [b : l]\, x_b^a\Big)^2 = m_l^2 \quad (l \in |G|) \tag{14-4}$$

Since (12-5) becomes trivial under (14-1), (14-2) and (14-4) are equivalent to the Landau equations. If we solve (14-2) and substitute x_b^a in (14-4), then according to Theorem 9-5 the left-hand side of (14-4) becomes Y_l^2 in the vertex representation.

After the above preliminaries, we consider the Feynman integral in the position space. By changing the parameters β_l into $\beta_l' = \beta_l/2$ in (8-6) and omitting the primes, (8-7) yields

$$f_G(p) = \text{const} \prod_{b \neq a} \left(\int d^4 x_b \right) \prod_l \left(\int_0^\infty d\beta_l \right) \exp\,[i\tilde{\Phi}(p, x, \beta) - \varepsilon] \qquad (14\text{-}5)$$

where $\varepsilon \equiv \frac{1}{2}\sum \varepsilon_l \beta_l^{-1}$ and

$$\tilde{\Phi}(p, x, \beta) \equiv \sum_{b \neq a} p_b x_b^a - \tfrac{1}{2} \sum_l \left\{ \beta_l \Big(\sum_{b \neq a} [b : l]\, x_b^a \Big)^2 + m_l^2/\beta_l \right\} \qquad (14\text{-}6)$$

Since the differentiations with respect to p_b simply yield a monomial of x_b^a in the integrand of (14-5), any derivative of $f_G(p)$ will not in general diverge because of the rapid oscillation of the integrand. Hence a real singularity for p real can occur only when the phase factor is stationary, that is,

$$\partial\tilde{\Phi}/\partial x_b^a = 0 \quad (b \neq a) \qquad (14\text{-}7)$$

$$\partial\tilde{\Phi}/\partial \beta_l = 0 \quad (l \in |G|) \qquad (14\text{-}8)$$

apart from the cases in which some of β_l become infinite. It is straightforward to see that (14-7) and (14-8) are equivalent to (14-2) and (14-4), respectively. Since $\tilde{\Phi}$ has the dimension of action, it should be replaced by $\tilde{\Phi}/\hbar$ if we do not employ natural units. If we take the classical limit $\hbar \to 0$, the only surviving contribution comes out from the stationary point, which satisfies (14-7) and (14-8). Thus the classical limit of the Feynman integral precisely yields the condition for the (leading) physical-region singularity, and x_b^a is the space-time position relative to x_a by definition. Therefore, Theorem 4-1 holds without making the *ad hoc* assumption $\alpha_l = \tau_l/m_l$.

14-2 Discontinuity Formula

Since most of the singularities of $f_G(p)$ are branch points, it is important to know the discontinuities around them. The best way of obtaining the discontinuity formula of $f_G(p)$ is the homological method [13, 14], but we do not describe it here.

The discontinuity around a real singularity is purely imaginary; $2i\,\text{Im} f_G(s)$ is a superposition of discontinuities. In order to extract a discontinuity around a single real singularity, we have to compute the *nonanalytic change* of $\text{Im} f_G(s)$ at that point $s = \bar{s}$, which is a difference between $\text{Im} f_G(s)$ for

$s > \bar{s}$ and the analytic continuation to $s > \bar{s}$ of $\operatorname{Im} f_G(s)$ for $s < \bar{s}$. The non-analytic change is easy to be found only when $\operatorname{Im} f_G(s) = 0$ for $s < \bar{s}$.

As remarked in Subsection 7-3, the Feynman integral satisfies the unitarity condition by itself. The discontinuity around a normal threshold corresponding to an intermediate state S can be seen to be given by $f_G^{(S)}(p)$ in the physical region [see (7-65)]. Therefore the expression for $f_G^{(S)}(p)$ can be regarded as a discontinuity formula. The *Cutkosky rule* [9] is a formal generalization of it to other singularities. The original prescription was as follows: The discontinuity around a Landau singularity corresponding to a reduced graph R should be given by an integral which is obtained from the Feynman integral by replacing the propagator (6-15) by $2\pi\delta_p(m_l^2 - k_l^2)$ for every $l \in |R|$, where δ_p is called a "proper" δ-function. When external momenta p are complex, the meaning of this rule is ambiguous because the contours of $\hat{k}_l$ are not specified. For physical-region singularities, however, the above rule is well defined if we set

$$\delta_p(m_l^2 - k_l^2) = \theta(k_l^{(0)})\,\delta(m_l^2 - k_l^2) \tag{14-9}$$

The validity of the Cutkosky rule for the physical-region singularity can be verified by the homological method [23, 24].

According to Pham [23], the Cutkosky rule for the physical-region singularity can be regarded as a more fundamental rule than the unitarity of the S-matrix. The S-matrix satisfies a kind of macrocausality condition, that is, if several elementary processes occur in very large spacelike separations, the S-matrix for the total process is a product of the S-matrices for those subprocesses. This property is called the *cluster property*. Since the physical-region singularity can be understood in terms of multiple scattering, it may be natural to call the Cutkosky rule a "cluster property for very large timelike separations".

14-3 Infrared Divergence

The infrared divergence, which is well known in quantum electrodynamics, can be regarded as a pathological case of the physical-region singularity [25]. Because of the 4-momentum conservation (12-4) and the mass-shell equations (12-7), the reduced graph R corresponding to a physical-region singularity cannot involve any vertex of F-degree 3 if no unstable particles are involved and if no massless particles are considered. Only in this subsection, however, we admit the presence of massless particles. Then vertices of F-degree 3 are of only one type, that is, R can involve a vertex b if two lines l and l' incident with b has the same mass m and the remaining one line k has zero mass. The Landau equations can be satisfied only if $[b:l]\,q_l + [b:l']\,q_{l'} = 0$, $q_k = 0$, and $q_l^2 = m^2$. We call such a vertex an *infravertex*. It is convenient to use the terminology of infravertex also for any vertex of F-degree $d \neq 3$ if two lines incident with it have the same mass m (m may equal zero) and if all the other lines incident with it have zero mass. The *infrared singularity* is defined to be the physical-region singularity for a reduced graph in which no pair of non-infravertices is linked by nonzero-mass line alone. For practical applications, we always consider infrared singularities for reduced graphs, but for simplicity we here consider only the leading infrared singularity as in Section 12.

As seen from the definition of the infrared singularity, the set of all lines which have nonzero mass is a disjoint union of paths and circuits which pass through no common vertices except for non-infravertices. Those paths and circuits are called *infrapaths* and *infracircuits*, respectively. More precisely, an infrapath is a path consisting of nonzero-mass lines either between two external vertices or between one external vertex and one non-infravertex. By definition, the masses on any infrapath or infracircuit are equal to each other.

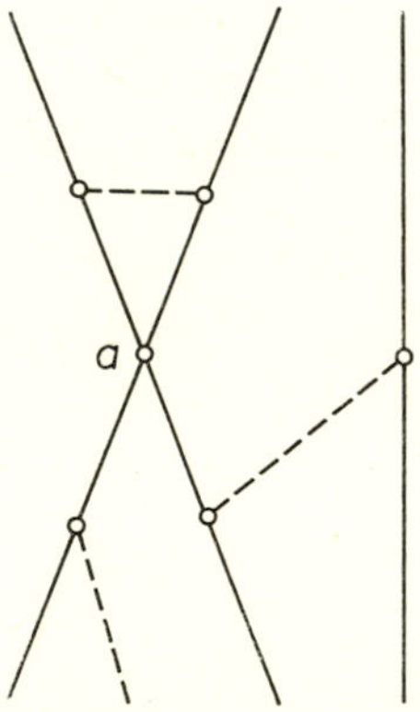

Fig. 14-2 A Feynman graph corresponding to infrared singularity. Dotted lines indicate massless-particle lines, and solid ones are of nonzero-mass particles. The vertex a is a non-infravertex.

Now, we consider the Landau equations. As noted above, so as to satisfy (12-4) and (12-7), the momenta q_l are equal to each other for all l belonging to any infrapath P or infracircuit C if the orientation of l is chosen so as to coincide with that of P or C, and $q_k = 0$ for all lines k of zero mass

($m_k = 0$). Next we consider the circuit equations (12-6). If C is an infracircuit, (12-6) implies $\sum_{l \in C} \alpha_l = 0$ and therefore $\alpha_l = 0$ for all $l \in C$ under the condition $\alpha_l \geqq 0$. Likewise, if C is an arbitrary circuit, which intersects m infrapaths $P_1, \ldots, P_m$, then (12-6) implies

$$\sum_{j=1}^{m} \Big(\sum_{l \in C \cap P_j} \alpha_l \Big) (\pm q_{P_j}) = 0 \tag{14-10}$$

where q_{P_j} is the momentum common to all lines of P_j, and the double sign depends only on j. If we assume that $q_{P_1}, \ldots, q_{P_m}$ are linearly independent because they are equal to momenta of distinct external particles,† then (14-10) implies $\sum_{l \in P \cap C} \alpha_l = 0$, that is, $\alpha_l = 0$ for all $l \in P \cap C$, where P is any infrapath. Thus we have $\alpha_l = 0$ for all (noncut) lines l of nonzero mass. The value $\alpha_l = 0$ should not be identified with the end-point value, but it should be regarded as a limiting case of $\alpha_l > 0$. This feature is a pathological aspect of the infrared singularity.

It is important to note that (12-6) cannot determine α_k if k is a line of zero mass because $q_k = 0$. If G involves ν lines of zero mass, $\nu - 1$ Feynman parameters are undetermined (the minus one is due to $\sum_{l \in |G|} \alpha_l = 1$). Therefore, when we calculate the behavior near the singularity, $\nu - 1$ coefficients b_j in (12-20) should vanish. Hence, the contribution from the leading infrared singularity‡ behaves like (12-22) with

$$\begin{aligned} k &= \tfrac{1}{2}(4\mu - N - 1) - \tfrac{1}{2}(\nu - 1) \\ &= \tfrac{1}{2}[4\mu - (N - \nu) - 2\nu] \end{aligned} \tag{14-11}$$

where ΔV should be proportional to a fictitious mass squared of the massless particles. If $k \leqq 0$ then the Feynman integral is divergent at the infrared singularity. This divergence is known as *infrared divergence*.§ As seen from (14-11), the occurrence of infrared divergence can easily be judged by counting the order of zero in the denominators of (6-23) [26].

† Here we need not consider C which intersects more than two infrapaths incident with a common non-infravertex.

‡ Here we neglect the contributions from non-leading infrared singularities.

§ The infrared divergence is usually the divergence in the case in which all infrapaths are incident with one non-infravertex, because only in this case the singularity becomes independent of all invariant squares of momenta other than external masses squared. The divergence in the case in which we have no non-infravertex and two infrapaths is known as the Coulomb-forward-scattering infinity.

§ 15 EXISTENCE REGION OF SINGULARITIES

Landau singularities of a Feynman function $f_G(p)$ can be present on the surfaces defined by the Landau equations. When internal masses m_l $(l \in |G|)$ vary in a certain region, say, in

$$m_l^2 \geqq 0 \quad (l \in |G|) \tag{15-1}$$

those surfaces sweep some region in the complex p space. A Feynman function with arbitrary internal masses satisfying (15-1) is holomorphic outside that region (apart from the possible second-type singularities described in the next section). To find the existence region of singularities is closely related to the problem of discussing the analyticity in the axiomatic field theory.

For the moment, we regard the Feynman function as a function of only one invariant s by holding all others fixed.

THEOREM **15-1** Let $s = \bar{s}$ and $\alpha = \bar{\alpha}$ be a (complex) solution of the Landau equations for the leading Landau singularity. When m_l^2 varies, the variation of $\bar{s}$ is given by

$$\frac{\partial \bar{s}}{\partial m_l^2} = -\frac{\bar{\alpha}_l}{\partial V/\partial s|_{\alpha=\bar{\alpha}}} \tag{15-2}$$

where V is considered as a function of s, α_l $(l \in |G|)$, and m_l^2 $(l \in |G|)$.

Proof By differentiating $V(\bar{s}, \bar{\alpha}, m^2)$ by m_l^2, we have

$$\frac{\partial V}{\partial m_l^2} = \bar{\alpha}_l + \frac{\partial V}{\partial \bar{s}} \frac{\partial \bar{s}}{\partial m_l^2} + \sum_{k \in |G|} \frac{\partial V}{\partial \bar{\alpha}_k} \frac{\partial \bar{\alpha}_k}{\partial m_l^2} \tag{15-3}$$

Since the variations of $\bar{s}$ and $\bar{\alpha}_l$ $(l \in |G|)$ occur in such a way that the Landau equations are still satisfied, we have

$$\frac{\partial V}{\partial m_l^2} = 0, \quad \frac{\partial V}{\partial \bar{\alpha}_l} = 0 \quad (l \in |G|) \tag{15-4}$$

On substituting (15-4) in (15-3), we find

$$\frac{\partial \bar{s}}{\partial m_l^2} = -\frac{\bar{\alpha}_l}{(\partial/\partial \bar{s})\, V(\bar{s}, \bar{\alpha}, m^2)} \tag{15-5}$$

a result which is identical with (15-2) because $V(s, \alpha, m^2)$ is linear with respect to s. q.e.d.

An important consequence of Theorem 15-1 is that $\bar{\alpha}_l^{-1}\,\partial\bar{s}/\partial m_l^2$ is independent of l [27].†

THEOREM **15-2** When internal masses vary in (15-1), the boundary of the singularity region is part of the union of curves Γ_I over all $I \subset |G|$. Here Γ_I is defined by $s = \bar{s}$ and the equations

$$m_l^2 = 0 \qquad \text{for all } l \in I \tag{15-6}$$

$$\partial\bar{s}/\partial m_l^2 = r_l\zeta \quad \text{for all } l \in I^* \tag{15-7}$$

where $I^* \equiv |G| - I$, r_l is real, and ζ is complex but independent of l.

We omit the proof of the above theorem; here we merely note that (15-7) can be obtained by rewriting the well-known condition for the envelope of a family of curves into that in the complex plane [28].

For $l \in I$, the quantity $r_l \equiv \zeta^{-1}\partial\bar{s}/\partial m_l^2$ is of course complex, but the signs of Im r_l are common for all $l \in I$. This is because the change of $\bar{s}$ for arbitrary variations of internal masses is written as

$$\begin{aligned} d\bar{s} &= \sum_{l\in|G|} (\partial\bar{s}/\partial m_l^2)\, dm_l^2 \\ &= \zeta\Big(\sum_{l\in I^*} r_l\, dm_l^2 + \sum_{l\in I} r_l\, dm_l^2\Big) \end{aligned} \tag{15-8}$$

with Im $r_l = 0$ for all $l \in I^*$ and $dm_l^2 \geqq 0$ for all $l \in I$, and Im $(d\bar{s}/\zeta)$ should have a definite sign since the singularity region lies in one side of its boundary. From Theorems 15-1 and 15-2, we obtain the following theorem.

THEOREM **15-3** [27] On the boundary of the singularity region, the Feynman parameters $\bar{\alpha}_l$ are proportional to r_l. Therefore, Im $\bar{\alpha}_l = 0$ for $l \in I^*$ and Im $\bar{\alpha}_l > 0$ for $l \in I$ if we neglect a common factor.

This theorem simplifies the analysis of the singularity region considerably. It is still prohibitively difficult, however, to analyze the general Feynman graph in detail. Fortunately, the singularity region seems not to vary with Feynman graphs very much. It seems natural to conjecture [29]‡ that all Landau singularities of the general Feynman graph having n external

† The original proof of this fact was more complicated than the one presented here.

‡ There is also a conjecture [30] that the n-point Wightman function, which is a position-space analytic function, has a singularity region whose boundary would be characterized by that of the complete Feynman graph having n external vertices. It seems, however, that this conjecture is groundless for $n \geqq 4$ [31, 32].

vertices would lie in the singularity region defined by the internal-mass superposition of the Landau singularities of the *complete* Feynman graph having n external vertices ($v(G) = g$). Hence it is important to analyze the complete Feynman graph.

For this purpose, it is convenient to employ the modified form, (14-2) and (14-4), of the Landau equations. In the matrix notation, (14-2) is rewritten as (see Subsection 8-1)

$$\mathscr{W}^a \mathbf{x}^a - \mathbf{p}^a = 0 \tag{15-9}$$

Hence

$$\mathbf{p}^a(\mathbf{p}^a)^t = \mathscr{W}^a \mathbf{x}^a (\mathbf{x}^a)^t \mathscr{W}^a \tag{15-10}$$

In a complete graph G, since every pair of distinct vertices uniquely correspond to a line, it is convenient to denote every line l by the pair of its end vertices b and c, and correspondingly we rewrite β_l and m_l as $\beta_{bc}(= \beta_{cb})$ and $m_{bc}(= m_{cb})$, respectively. Then (14-4) becomes

$$\begin{aligned} (x_b^a - x_c^a)^2 &= m_{bc}^2/\beta_{bc}^2 \quad (b, c \in g - a) \\ x_b^a &= m_{ab}^2/\beta_{ab}^2 \quad (b \in g - a) \end{aligned} \tag{15-11}$$

with $m_{bb}^2 \equiv 0$. Hence

$$x_b^a x_c^a = \frac{1}{2}\left(\frac{m_{ab}^2}{\beta_{ab}^2} + \frac{m_{ac}^2}{\beta_{ac}^2} - \frac{m_{bc}^2}{\beta_{bc}^2}\right) \quad (b, c \in g - a) \tag{15-12}$$

Let $\mathscr{M}^a$ be the square matrix of order $n - 1$ whose elements are defined by the right-hand side of (15-12); then

$$\mathbf{x}^a(\mathbf{x}^a)^t = \mathscr{M}^a \tag{15-13}$$

On substituting (15-13) in (15-10), we obtain

$$\mathbf{p}^a(\mathbf{p}^a)^t = \mathscr{W}^a \mathscr{M}^a \mathscr{W}^a \tag{15-14}$$

In (15-14), $n(n - 1)/2$ scalar product of external momenta are expressed in terms of $n(n - 1)/2 - 1$ auxiliary variables $\beta_{bc}/\sum_{b' \neq c'} \beta_{b'c'}$ together with $n(n - 1)/2$ mass parameters.

Now, we apply the theorems given above to our complete Feynman graph. There are several nontrivial choices of I, but the most interesting case is the case in which I is a star. Hence we put $I = S[a]$. Then Theorems 15-2 and 15-3 (remember that β_l is inversely proportional to $\bar{\alpha}_l$) lead us to

$$\begin{aligned} m_{ab}^2 = 0, \quad \operatorname{Im} \beta_{ab} &< 0 \quad \text{for } b \in g - a \\ \operatorname{Im} \beta_{bc} &= 0 \quad \text{for } b, c \in g - a \end{aligned} \tag{15-15}$$

where we have redefined β_{bc} by dividing it by a single complex factor. On account of (15-15), we have

$$\begin{aligned}\mathscr{W}^a &= \Big(\sum_{l\in|G|} [b:l]\,[c:l]\,\beta_l \,\big|\, b, c \in g - a\Big) \\ &= \Big(\sum_{d\in g} \beta_{db}\delta_{bc} - \beta_{bc} \,\big|\, b, c \in g - a\Big) \end{aligned} \qquad (15\text{–}16)$$

$$\mathscr{M}^a = -\tfrac{1}{2}(m_{bc}^2/\beta_{bc}^2 \mid b, c \in g - a) \qquad (15\text{-}17)$$

From (15-16) and (15-15)

$$\operatorname{Im} \mathscr{W}^a = (\operatorname{Im} \beta_{ab} \cdot \delta_{bc} \mid b, c \in g - a) \qquad (15\text{-}18)$$

Thus the matrix $\mathscr{W}^a$ is symmetric, and its imaginary part is diagonal and negative definite; $\mathscr{M}^a$ is real and symmetric and its diagonal elements are zero. Therefore, $\boldsymbol{p}^a(\boldsymbol{p}^a)^t$ is expressed in the *Jost canonical form*, usually called DANAD, *except for the overall sign* [30].

In the axiomatic field theory, DANAD appears as the boundary of the singularity region of a position-space analytic function obtained without using the microcausality condition (6-12). Apart from the trivial case $n = 2$, the implication of (6-12) was fully utilized only when $n = 3$ [33]. The main part of the boundary surface found in this case exactly coincides with the boundary of the singularity region for the $n = 3$ complete Feynman graph obtained above.

16-1 Pure Second-Type Singularities

The integrand of a Feynman-parametric integral has two denominators U and V. If $\alpha \in \Delta$, U can vanish only at the end points. If one considers complex singularities of a Feynman function on unphysical sheets, however, U can vanish for complex nonzero values of Feynman parameters. The singularity caused by a complex zero point of U is called a *second-type singularity*.† A characteristic feature of a second-type singularity is that its location is (at least partially) independent of internal masses. If it is completely independent of them, the singularity is called a *pure* second-type singularity; otherwise it is called a *mixed* second-type singularity. The former is discussed in this subsection.

It is known that the existence of second-type singularities depends on the dimension ν of the momentum space. By extending (11-4) to the ν-dimensional case (see (7-72)), we have

$$f_G(p) = \text{const} \int_0^1 d\alpha_0 \int_\Delta d^{N-1}\alpha \frac{\alpha_0^{r-1}(1-\alpha_0)^{N-\frac{1}{2}\nu\mu-1}}{U^{\nu/2}[\alpha_0\Lambda + (1-\alpha_0)V]^{N-\frac{1}{2}\nu\mu+r}} \tag{16-1}$$

where $N - \frac{1}{2}\nu\mu + r > 0$ and $r > 0$. It is convenient to transform the integration variables by $(1-\alpha_0)\,\alpha_l = \alpha_l'\,(l \in |G|)$. After dropping the primes of the new variables, we obtain

$$f_G(p) = \text{const} \int_{\Delta'} d^N\alpha \frac{\alpha_0^{r-1}}{U^{\nu/2}V'^{N-\frac{1}{2}\nu\mu+r}} \tag{16-2}$$

where

$$\int_{\Delta'} d^N\alpha \equiv \int_0^\infty d\alpha_0 \prod_{l\in|G|}\left(\int_0^\infty d\alpha_l\right)\delta\left(1-\alpha_0-\sum_l \alpha_l\right) \tag{16-3}$$

and

$$V' \equiv \alpha_0\Lambda + V \quad (\Lambda > 0) \tag{16-4}$$

The use of the parameter α_0 is suitable for the investigation of the second-type singularity.

† The existence of the second-type singularities was found by Cutkosky [9]. Their systematic investigation was made first by Fairlie, Landshoff, Nuttall, and Polkinghorne [34].

Since r can be arbitrarily large, we may choose it in such a way that

$$m \equiv N - \tfrac{1}{2}\nu(\mu + 1) + r \tag{16-5}$$

is a nonnegative integer. Then (16-2) is rewritten as

$$f_G(p) = \text{const} \int_{\Delta'} d^N\alpha \frac{\alpha_0^{r-1}\, U^m}{(UV')^{\frac{1}{2}\nu+m}} \tag{16-6}$$

Since the numerator of its integrand is a polynomial, we have only to investigate the denominator. Hence, as in Section 12, the leading singularity is given by

$$(\partial/\partial\alpha_0)\, UV' = 0 \tag{16-7}$$

$$(\partial/\partial\alpha_l)\, UV' = 0 \quad (l \in |G|) \tag{16-8}$$

Since $\partial U/\partial\alpha_0 = 0$ and $\partial V'/\partial\alpha_0 = \Lambda \neq 0$, (16-7) reduces to

$$U = 0 \tag{16-9}$$

Thus it characterizes the second-type singularity. On the contrary, if $\alpha_0 = 0$ instead of (16-7), (16-8) reduces to the equations which determine a Landau singularity.

In order to solve (16-8) with (16-9) for $\alpha_0 \neq 0$, it is convenient to employ (7-22), that is,

$$UV = UL + (\hat{\boldsymbol{q}}_{T*})^t\, \mathscr{U}_{T*}^{\mathrm{adj}}\, \hat{\boldsymbol{q}}_{T*} \tag{16-10}$$

where

$$\mathscr{U}_{T*}^{\mathrm{adj}} \equiv ([\mathscr{U}_{T*}]^{l,l'} \mid l, l' \in T^*) \tag{16-11}$$

and L is defined by (7-10). By definition,

$$\mathscr{U}_{T*}\mathscr{U}_{T*}^{\mathrm{adj}} \equiv \mathscr{U}_{T*}^{\mathrm{adj}}\mathscr{U}_{T*} \equiv U \cdot \mathscr{E} \tag{16-12}$$

where $\mathscr{E}$ denotes the unit matrix of order μ. Suppose that the internal momenta q_l ($l \in |G|$) are expressed as linear combinations of external momenta p_b ($b \neq a$). Then we can write

$$\hat{\boldsymbol{q}}_{T*} = \mathscr{K}\boldsymbol{p}^a \tag{16-13}$$

where $\mathscr{K}$ is a certain $\mu \times (n - 1)$ matrix depending on Feynman parameters. If there exists a lightlike vector which is orthogonal to p_b for all $b \in g$, we denote it by ϱ; if not so, let $\varrho \equiv 0$. Then we make an *ansatz*

$$\hat{\boldsymbol{q}}_{T*} = \mathscr{U}_{T*}\tilde{\boldsymbol{q}} + \varrho \boldsymbol{e} \tag{16-14}$$

where $\tilde{\boldsymbol{q}}$ is a column vector consisting of μ momenta orthogonal to ϱ, and $\boldsymbol{e}$ is a column vector consisting of μ arbitrary numbers. Since $\mathscr{U}_{T*}$ is singular in the present case, we can find a column vector $\boldsymbol{r}$ consisting of μ numbers such that $\mathscr{U}_{T*}\boldsymbol{r} = 0$. [If we assume rank $(\mathscr{U}_{T*}) = \mu - 1$, $\boldsymbol{r}$ is unique apart from a multiplicative factor.] Then $\tilde{\boldsymbol{q}}$ exists only if

$$\boldsymbol{r}^t\hat{\boldsymbol{q}}_{T*} = \varrho\boldsymbol{r}^t\boldsymbol{e} \tag{16-15}$$

From (16-13) and (16-15), we see that a certain linear combination of external momenta is equal to a lightlike vector which is orthogonal to them. By multiplying this equation by $p_b\,(b \in g - a)$ and eliminating the coefficients of the scalar products $p_b p_c$, we have the Gram-determinant equation

$$\det\,[p^a(p^a)^t] = 0 \tag{16-16}$$

which is the characteristic equation of the pure second-type singularity [34].

Now, we substitute (16-14) in (16-10) and differentiate it with respect to α_l. By using (16-12) and (16-9), we have

$$\frac{\partial}{\partial\alpha_l}UV = L\frac{\partial U}{\partial\alpha_l} + \tilde{\boldsymbol{q}}^t\mathscr{U}_{T*}\frac{\partial\mathscr{U}_{T*}^{\text{adj}}}{\partial\alpha_l}\mathscr{U}_{T*}\tilde{\boldsymbol{q}} \tag{16-17}$$

Since (16-12) implies

$$\mathscr{U}_{T*}\frac{\partial\mathscr{U}_{T*}^{\text{adj}}}{\partial\alpha_l} + \frac{\partial\mathscr{U}_{T*}}{\partial\alpha_l}\mathscr{U}_{T*}^{\text{adj}} = \frac{\partial U}{\partial\alpha_l}\mathscr{E} \tag{16-18}$$

(16-17) is rewritten as

$$\frac{\partial}{\partial\alpha_l}UV = \frac{\partial U}{\partial\alpha_l}(L + \tilde{\boldsymbol{q}}^t\mathscr{U}_{T*}\tilde{\boldsymbol{q}}) \tag{16-19}$$

Thus (16-7) and (16-8) reduce to (16-9), (16-14), and

$$\alpha_0\Lambda + L + \tilde{\boldsymbol{q}}^t\mathscr{U}_{T*}\tilde{\boldsymbol{q}} = 0 \tag{16-20}$$

The last equation (16-20) determines α_0, that is, it yields no restriction on α_l and $q_l\,(l \in |G|)$. Thus we need not solve $\tilde{\boldsymbol{q}}$ from (16-14).

Since the above method is based on the ansatz (16-14), it is instructive to analyze (16-8) more directly. By means of the modified cut-set representation (7-46), we have

$$\frac{\partial}{\partial\alpha_l}UV = \frac{\partial U}{\partial\alpha_l}\sum_k \alpha_k m_k^2 - \sum_{(h|g-h)}\frac{\partial W^{(h|g-h)}}{\partial\alpha_l}\Big(\sum_{b\in h} p_b\Big)^2 \tag{16-21}$$

with the aid of (16-9). On setting

$$\xi \equiv \alpha_0\Lambda + \sum_k \alpha_k m_k^2 \tag{16-22}$$

we find that (16-8) becomes

$$\sum_{(h|g-h)} \frac{\partial W^{(h|g-h)}}{\partial \alpha_l} \Big(\sum_{b \in h} p_b \Big)^2 = \xi \frac{\partial U}{\partial \alpha_l} \quad (l \in |G|) \tag{16-23}$$

This equation shows the reason why the pure second-type singularity is independent of internal masses.

We consider a single-loop graph having no internal vertex (i.e., $\mu = 1$ and $N = n$). We number the internal lines cyclically. Then

$$U = \sum_{j=1}^{n} \alpha_j \tag{16-24}$$

$$W^{(h|g-h)} = W_S = \alpha_j \alpha_k \text{ for } S = \{j, k\} \in S(h \mid g - h) \tag{16-25}$$

On writing $\sum_{b \in h} p_b = p_{jk} = -p_{kj}$, (16-9) and (16-23) become

$$\sum_{j=1}^{n} \alpha_j = 0 \tag{16-26}$$

$$\sum_{j \neq k} \alpha_j p_{jk}^2 = \xi \quad (k = 1, \dots, n) \tag{16-27}$$

respectively. By eliminating ξ and $\alpha_1, \dots, \alpha_n$, we find

$$\begin{vmatrix} 0 & 1 & 1 & \cdots & 1 \\ 1 & 0 & p_{12}^2 & \cdots & p_{1n}^2 \\ 1 & p_{21}^2 & 0 & \cdots & p_{2n}^2 \\ & \cdot & \cdot & \cdot & \cdot \\ 1 & p_{n1}^2 & p_{n2}^2 & \cdots & 0 \end{vmatrix} = 0 \tag{16-28}$$

a result which is equivalent to (16-16) because of a well-known identity between determinants, called the Caylay transformation.

When (16-28) is satisfied, the ratio of $\alpha_0, \alpha_1, \dots, \alpha_n$ is uniquely determined. Hence the behavior of $f_G(p)$ near the singularity is given by (12-22) with

$$k = -(\tfrac{1}{2}\nu + m) + m + \tfrac{1}{2}N = -\tfrac{1}{2}(\nu - N) \tag{16-29}$$

according to (16-6). It is known, however, that the second-type singularity is present only if $\nu > n(= N)$ in the single-loop graph [34]. Hence the behavior near the singularity is given by

$$(\Delta V)^{-(\nu - n)/2} \quad \text{for } \nu > n \tag{16-30}$$

In the general graph, the pure second-type singularity can be present only for $n \leqq \nu + 1$. This is because if $n > \nu + 1$ then the external momenta p

are always linearly dependent so that they identically satisfy (16-16), but it is impossible that $f_G(p)$ is singular for all values of p.

We note that the behavior near a second-type singularity becomes stronger if G involves particles with nonzero spins, because Y_l and $X_{ll'}$ contain U in their denominators. Thus the second-type singularity depends not only on the dimension of the space but also on the spins of particles.

Example **1** The $N = 2$ self-energy graph ($\mu = 1$, $n = 2$). Let $s = p^2$; (16-26) and (16-27) with (16-22) lead us to

$$\alpha_2 = -\alpha_1, \quad s = \xi = 0$$

$$\alpha_0 = \Lambda^{-1}\alpha_1 (m_2^2 - m_1^2)$$

The case $m_1 = m_2$ is exceptional because then $\alpha_0 = 0$ (a Landau singularity coincides with the second-type singularity); the behavior near the singularity is always like $s^{-\frac{1}{2}}$ independently of ν. For $m_1 \neq m_2$, (16-30) tells us that $f_G(p)$ behaves like $s^{-(\nu-2)/2}$. Indeed,

$$\operatorname{Im} f_G(p) = \text{const}\, s^{-(\nu-2)/2} \{[s - (m_1 - m_2)^2][s - (m_1 + m_2)^2]\}^{(\nu-3)/2} \times \theta(s - (m_1 + m_2)^2)$$

whence $f_G(p)$ is singular at $s = 0$ on an unphysical sheet, provided that $\nu \geqq 3$.

Example **2** $\nu = 4$, $\mu = 1$, $n = N$. The only second-type singularities are those of the above self-energy graph and of the triangle graph (Fig. 13-1) because of the condition $N < 4$. In the latter, $f_G(p)$ behaves like [9]

$$[p_1^2 p_2^2 - (p_1 p_2)^2]^{-\frac{1}{2}}$$

Example **3** $\mu = 2$, $N = 3$, $n = 2$ (Fig. 13-3). Since (16-23) is satisfied by $s = \xi = 0$, the ratio $\alpha_1 : \alpha_2 : \alpha_3$ is *not* uniquely determined as in the case of the infrared singularity. Correspondingly, N in (16-29) should be replaced by $N - 1 = 2$. Thus $f_G(p)$ behaves like $(\Delta V)^{-(\nu-2)/2}$. Indeed, for $\nu = 3$ we have

$$\operatorname{Im} f_G(p) = \text{const}\, s^{-\frac{1}{2}}(s^{\frac{1}{2}} - m_1 - m_2 - m_3)\, \theta(s - (m_1 + m_2 + m_3)^2)$$

16-2 Mixed Second-Type Singularities

The mixed second-type singularities appear when $\mu \geqq 2$. The simplest example of them is provided by the so-called icecream-cone graph ($\mu = 2$, $N = 4$, $n = 3$) [34, 35].

The ansatz (16-14) in Subsection 16-1 has been made under the implicit assumption that all Feynman parameters are of the same order of magnitude. If some of them tend to zero as $U \to 0$, then we can find another type of solutions to (16-8), which are nothing but mixed second-type singularities.

Let H be a subgraph such that $|H|$ is a union of independent circuits ($\mu(H) > 0$). Suppose that

$$\delta \equiv \sum_{l \in |H|} |\alpha_l| \tag{16-31}$$

is infinitesimal and that $U \sim \delta^{\mu(H)+1}$ so that $U/\delta^{\mu(H)} \to 0$ as $\delta \to 0$. Then, since $U_R \nrightarrow 0$ as $\delta \to 0$, where $R \equiv G/|H|$, we find

$$U_H \sim \delta^{\mu(H)+1} \tag{16-32}$$

because of (9-1), that is, $U_H/\delta^{\mu(H)} \to 0$ as $\delta \to 0$.

Now, we consider (16-8). By using the vertex representation (7-45) of V, (16-8) is written as

$$Um_l^2 + \frac{\partial U}{\partial \alpha_l}\xi + \sum_{\{b,c\} \subset g} \frac{\partial W^{(b|c)}}{\partial \alpha_l} p_b p_c = 0 \quad (l \in |G|) \tag{16-33}$$

where ξ is defined by (16-22). For $l \in |H|$, the leading order ($\sim \delta^{\mu(H)-1}$) becomes independent of l. Indeed, we obtain

$$U_R \xi + \sum_{\{b,c\} \subset g} W_R^{(b|c)} p_b p_c = 0 \tag{16-34}$$

with the aid of (9-1) and (9-6). Therefore, in order to obtain $N(H) - 1$ independent equations, we should consider the next order ($\sim \delta^{\mu(H)}$). Because $U \sim \delta^{\mu(H)+1}$, those equations are independent of m_l ($l \in |H|$). On the other hand, for $l \in |R|$, the leading order is $\delta^{\mu(H)+1}$ because of (16-32) together with (9-1) and (9-6). Hence those $N(R)$ equations of (16-33) are dependent on m_l ($l \in |R|$) in the leading order.

Thus the mixed second-type singularity depends only on the internal masses of R.

References

(1) R. J. Eden, *Proc. Roy. Soc.* A **210**, 388 (1952).
(2) R. Karplus, C. M. Sommerfield, and E. H. Wichmann, *Phys. Rev.* **111**, 1187 (1958).
(3) Y. Nambu, *Nuovo Cim.* **9**, 610 (1958).
(4) R. Oehme, *Phys. Rev.* **111**, 1430 (1958).
(5) L. D. Landau, *Nucl. Phys.* **13**, 181 (1959).
(6) N. Nakanishi, *Progr. Theoret. Phys.* **22**, 128 (1959).

(7) J. D. Bjorken, unpublished (thesis).
(8) R. J. Eden, P. V. Landshoff, D. I. Olive, and J. C. Polkinghorne, *The Analytic S-Matrix* (Cambridge University Press, London, 1966).
(9) R. E. Cutkosky, *J. Math. Phys.* **1**, 429 (1960).
(10) D. Fotiadi, M. Froissart, J. Lascoux, and F. Pham, *Topology* **4**, 159 (1965).
(11) H. Araki, *Progr. Theoret. Phys. Suppl.* **18**, 83 (1961).
(12) D. Hall and A. S. Wightman, *Mat. Fys. Medd. Dan. Vid. Selsk.* **31**, No. 5 (1957).
(13) R. C. Hwa and V. L. Teplitz, *Homology and Feynman Integrals* (W. A. Benjamin Inc., N. Y., 1966).
(14) F. Pham, *Introduction à L'étude Topologique de Singularités de Landau* (Gauthier-Villars, Paris, 1967).
(15) J. B. Boyling, *Nuovo Cim.* **53**A, 351 (1968).
(16) N. Nakanishi, *Progr. Theoret. Phys. Suppl.* **18**, 1 (1961); Errata, *Progr. Theoret. Phys.* **26**, 806 (1961).
(17) J. Mathews, *Phys. Rev.* **113**, 381 (1959).
(18) N. Nakanishi, *Progr. Theoret. Phys.* **23**, 284 (1960).
(19) J. M. Westwater, *Helv. Phys. Acta* **40**, 596 (1967).
(20) S. Coleman and R. E. Norton, *Nuovo Cim.* **38**, 438 (1965).
(21) R. Karplus, C. M. Sommerfield, and E. H. Wichmann, *Phys. Rev.* **114**, 376 (1959).
(22) N. Nakanishi, *Progr. Theoret. Phys.* **39**, 768 (1968).
(23) F. Pham, *Annales de l'Institute Henri Poincaré* A **6**, 89 (1967).
(24) J. B. Boyling, *Nuovo Cim.* **44**, 379 (1966).
(25) T. Kinoshita, *J. Math. Phys.* **3**, 650 (1962).
(26) N. Nakanishi, *Progr. Theoret. Phys.* **19**, 159 (1958).
(27) B. Andersson, *Nucl. Phys.* **74**, 601 (1965).
(28) G. Källén, *Nucl. Phys.* **25**, 568 (1961).
(29) N. Nakanishi, *J. Math. Phys.* **3**, 1139 (1962).
(30) A. C. T. Wu, *Phys. Rev.* **135**, 222 (1964).
(31) M. Minami and N. Nakanishi, *Progr. Theoret. Phys.* **40**, 167 (1968).
(32) N. Nakanishi, *Progr. Theoret. Phys.* **42**, 966 (1969).
(33) G. Källén and A. Wightman, *Mat. Fys. skr. Dan. Vid. Selsk.* **1**, No. 6 (1958).
(34) D. B. Fairlie, P. V. Landshoff, J. Nuttall, and J. C. Polkinghorne, *J. Math. Phys.* **3** 594 (1962).
(35) I. T. Drummond, *Nuovo Cim.* **29**, 720 (1963).

CHAPTER 4

Perturbation-Theoretical Integral Representations

The extended transition amplitude is a sum of the Feynman integrals which are generated by a particular Lagrangian. It is therefore important to find a domain in which all Feynman functions corresponding to a certain set of Feynman graphs are holomorphic. This approach is complementary to the one developed in Chapter 3.

Without analyzing the explicit dependence on the Feynman parameters, we can obtain a certain amount of knowledge about such a domain as mentioned above from the Feynman-parametric formula. For this purpose, it is convenient to introduce the perturbation-theoretical integral representation (PTIR), which is obtained from the Feynman-parametric integral by changing integration variables in such a way that the number of new variables is equal to that of independent invariant squares of momenta. The most essential problem is to find support properties of the weight distributions of PTIR.

If the Feynman function is regarded as a function of only one invariant square, PTIR reduces to a (single) spectral representation, which is called a dispersion relation when applied to the scattering amplitude. Aside from this case, PTIR (a very special case of it) was used first by Wick [1] and Cutkosky [2] in 1954 in order to solve the Bethe-Salpeter equation for bound states. At the end of 1954, Nambu [3] proposed an integral representation for the scattering amplitude on the basis of Feynman integrals. In 1955, he [4] further proposed some multiple spectral representations, which unfortunately turned out to be wrong. In 1958, Mandelstam [5] proposed a double spectral representation for the scattering amplitude, called the Mandelstam representation. On the other hand, an integral representation for the vertex function†, which was a special case of PTIR, was derived, but

† This representation had, already in 1957, been applied to the Bethe–Salpeter equation as a natural generalization of the Wick–Cuskosky one [6].

unfortunately incorrectly, on the basis of the axiomatic field theory independently in 1959 by Faǐnberg [7], by Deser, Gilbert, and Sudarshan [8], and later by Ida [9]. The systematic proposal and detailed study of PTIR were made by Nakanishi [10-16] in 1961–1964. He also investigated support properties for various cases, and in particular proved the validity of the above-mentioned integral representation (under the stability condition) for any Feynman integral [10]. The PTIR for the scattering amplitude is quite akin to the Mandelstam representation in its appearance. Some important improvements of support properties were made by Boyling [17].

In this chapter, we develop the theory of PTIR. We derive PTIR from the Feynman-parametric formula in Section 17. Mathematical properties of PTIR are presented in Section 18. Methods of finding support properties are discussed in Sections 19–21; Section 19 is an application of the transport problem (see Section 5) and Section 21 offers an interesting graph-theoretical problem. In Section 22, we summarize the results on the support properties of PTIR.

§ 17 DEFINITION OF PTIR

17-1 General Remarks

As is well known in hadron physics, the dispersion relation is one of the most powerful tools for the analysis of strong interaction because it involves no approximation. Here a *dispersion relation* is a spectral representation of a scattering amplitude in a *single* invariant square (more precisely, it consists of two terms each of which has a spectral representation). Since the scattering amplitude is a function of two variables, however, the dispersion relation becomes unsuitable for investigating detailed dynamical problems. We shall need analyticity in two or more variables. In this direction, the Mandelstam representation† is the most interesting hypothesis for the two-variable analyticity of the scattering amplitude. Unfortunately, however, this representation is not yet proven even in the case of no anomalous threshold, and it is hopeless to extend it to the case of the production amplitudes. It may be worthwhile, therefore, to investigate a family of integral representations which are valid for various transition amplitudes and extended ones.

As described in Subsection 6-1, an extended transition amplitude is expressed in terms of an infinite sum of the Feynman integrals which are generated by a particular Lagrangian. Therefore, its analyticity domain can be obtained by investigating the analyticity domains of all those Feynman integrals, aside from the convergence problem of the perturbation series (see Section 24). For this purpose, we introduce the perturbation-theoretical integral representation (PTIR) in the next subsection.

We consider certain (infinite) sets of Feynman graphs G having n external lines. Those external lines are incident with n external vertices which may *not* necessarily be distinct. For simplicity of notation, however, we use n different names for the external vertices throughout this chapter. [Therefore, the meaning of n is somewhat different from that in Chapter 3.] The set of all external vertices of G is denoted by g as before.

The set, Ψ_n, of (connected) Feynman graphs is characterized by the kinds of particles associated with n external lines and the admissible types of vertices, where the *type* of a vertex a is a list of the particles of the lines (including external lines) incident with a. Of course, as required physically, particles of the same kind should have the same mass. A more detailed specification of Ψ_n (or Ψ_n^r) is stated in Subsection 20-1.

† Its explicit form is presented in Subsection 18-2.

The above characterization of the set Ψ_n is based on the Lagrangian field theory because the interaction Lagrangian determines the admissible types of vertices. In the axiomatic field theory, however, one cannot specify the type of each vertex because the Lagrangian is not considered there, and therefore one uses the spectral conditions as its substitute. Accordingly, in order to see the connection with the axiomatic field theory, it is interesting to consider another set, Φ_n, of (connected) Feynman graphs specified by the spectral conditions. Here the *spectral conditions* are defined in terms of Feynman graphs as follows: There exists a nonnegative constant $M_h (\equiv M_{g-h})$ for each nontrivial division $(h \mid g - h)$ of g such that for any Feynman graph $G \in \Phi_n$ we have

$$\sum_{l \in S} m_l \geqq M_h \quad \text{for any } S \in S_G(h \mid g - h) \tag{17-1}$$

that is, the total mass of any intermediate state in the channel $(h \mid g - h)$ has a lower bound M_h. In particular, if $h = \{a\}$ then we write M_h as M_a; if $h = \{a, b\}$ then $M_h = M_{ab}$, etc.†

It is noteworthy that the set Ψ_n also satisfies the spectral conditions if we define

$$M_h \equiv \inf_{G \in \Psi_n} \min_{S \in S_G(h \mid g-h)} \sum_{l \in S} m_l \tag{17-2}$$

and that (17-2) can easily be computed if the admissible types of vertices are given. Thus, in this sense, the conditions imposed on Φ_n is weaker than those on Ψ_n. The distinction between Φ_n and Ψ_n becomes very important if we have, for example, the baryon-number conservation law, which implies the existence of baryon paths between pairs of baryon external lines (see Subsection 20-1 for details). There is no way of expressing such a property in the spectral conditions. Owing to such circumstances, the supports of the weight distributions of PTIR for Ψ_n become more restrictive than those for Φ_n, provided that they satisfy the same spectral conditions.

For a transition amplitude (on the mass shell), we usually require p_a for any $a \in g$ to satisfy the *stability condition*

$$p_a^2 \leqq M_a^2 \tag{17-3}$$

Throughout this chapter, we neglect spins, that is, all particles are regarded as spinless.

† We use this notation for any quantity having a suffix h.

17-2 Derivation of PTIR

In this subsection, we derive PTIR from the Feynman-parametric formula. We note that it contains external momenta only in the denominator function V as scalar products or invariant squares, and that V is a linear function of them. Since we often consider the quantities on the mass shell, it is convenient to employ invariant squares as variables. Hence the modified cut-set representation is suitable for our purpose.

The number of independent invariants, that of scalar products, and that of invariant squares vary with n as follows:

n	independent variables	scalar products	invariant squares
2	1	1	1
3	3	3	3
4	6	6	7
$n \geqq 5$	$4n - 10$	$n(n-1)/2$	$2^{n-1} - 1$

Since the Feynman function should be expressed as a function of independent invariants, we have to eliminate some of invariant squares for $n \geqq 4$. For $n = 4, 5$, all relations between squares are linear, and therefore it is not difficult to eliminate the redundant ones. For $n \geqq 6$, however, since the external momenta are no longer linearly independent, the Gram determinants for all sets of five momenta vanish identically, and they present nonlinear relations for squares. If one eliminates redundant squares by using them, then the linearity of V in squares is lost. Thus it is not convenient to employ squares for $n \geqq 6$ if our momentum space is of four dimensions. Hereafter, we do not consider the complication due to the dimension of the Minkowski space, that is, we neglect the non linear relations.

We denote the set of all independent squares by s symbolically. All functions of external momenta are regarded as functions of s. Hence we write (11-3) as

$$f_G(s, \lambda) = \text{const} \int_\Delta d^{N-1}\alpha \frac{\prod_l \alpha_l^{\lambda_l - 1}}{U^2(\alpha)\,[V(s, \alpha)]^{\lambda_0 - 2\mu}} \tag{17-4}$$

The modified cut-set representation (7-46) of V reads

$$V = \sum_l \alpha_l m_l^2 - \sum_h \zeta_h s_h \tag{17-5}$$

where

$$\zeta_h \equiv W^{(h|g-h)}/U \tag{17-6}$$

$$s_h \equiv \Big(\sum_{b\in h} p_b\Big)^2 \tag{17-7}$$

and the summation $\sum_h$ goes over all nontrivial divisions $(h \mid g - h)$ of g. Of course $\zeta_h \geqq 0$ for $\alpha \in \Delta$. We multiply (17-4) by

$$1 \equiv \prod_h \left[\int_0^1 \delta(z_h - \zeta_h/\xi)\, dz_h\right] \int_0^\infty d\varkappa\, \delta\Big(\varkappa - \sum_l \alpha_l m_l^2/\xi\Big) \tag{17-8}$$

where

$$\xi \equiv \sum_h \zeta_h \tag{17-9}$$

and interchange the order of the integrations. Then we obtain

$$f_G(s, \lambda) = \prod_h \left(\int_0^1 dz_h\right) \delta\Big(1 - \sum_h z_h\Big) \int_0^\infty d\varkappa \frac{\varphi_G(z, \varkappa; \lambda)}{\Big(\varkappa - \sum_h z_h s_h\Big)^{\lambda_0 - 2\mu}} \tag{17-10}$$

with

$$\delta\Big(1 - \sum_h z_h\Big) \varphi_G(z, \varkappa; \lambda) \equiv \text{const} \int_\Delta d^{N-1}\alpha \frac{\prod_l \alpha_l^{\lambda_l - 1}}{U^2 \cdot \xi^{\lambda_0 - 2\mu}}$$

$$\times \prod_h \delta(z_h - \zeta_h/\xi) \cdot \delta\Big(\varkappa - \sum_l \alpha_l m_l^2/\xi\Big) \tag{17-11}$$

Now, we perform the regularization defined by (10-31) and integrate (17-10) by parts with respect to $\varkappa$ $N - 2\mu - 1$ times. [If $N - 2\mu \leqq 0$ then we need subtraction procedure, which is described in Subsection 18-3. In this section, we neglect this complication.] Then we have

$$\bar{f}_G(s) = \prod_h \left(\int_0^1 dz_h\right) \delta\Big(1 - \sum_h z_h\Big) \int_0^\infty d\varkappa \frac{\bar{\varphi}_G(z, \varkappa)}{\varkappa - \sum_h z_h s_h} \tag{17-12}$$

where

$$\bar{\varphi}_G(z, \varkappa) \equiv (\partial/\partial\varkappa)^{N-2\mu-1} \operatorname*{Reg}_{\lambda=1} \varphi_G(z, \varkappa; \lambda)/(N - 2\mu - 1)! \tag{17-13}$$

It is noteworthy that in (17-12) the dependence on G of the integral is contained only in the weight distribution $\bar{\varphi}_G(z, \varkappa)$. Therefore we obtain an integral representation associated with the set Φ_n or Ψ_n if we replace $\bar{\varphi}_G(z, \varkappa)$

by a weight distribution $\varphi_n(z, \varkappa)$, which is a formal linear combination of $\bar{\varphi}_G(z, \varkappa)$ over all $G \in \Phi_n$ or Ψ_n. When $\bar{\varphi}_G(z, \varkappa)$ is replaced by $\varphi_n(z, \varkappa)$, the right-hand side of (17-12) is denoted by $f_n(s)$, that is,

$$f_n(s) = \prod_s \left(\int_0^1 dz_h \right) \delta\left(1 - \sum_h z_h\right) \int_0^\infty d\varkappa \frac{\varphi_n(z, \varkappa)}{\varkappa - \sum_h z_h s_h} \tag{17-14}$$

By definition, $f_n(s)$ is dependent only on Φ_n or Ψ_n but not on the individual Feynman graph. In this way, we can express some common properties of the Feynman functions associated with Φ_n or Ψ_n.

Since the invariant squares s_h are independent for $n = 2$ and $n = 3$, the representation (17-14) is meaningful in those cases. Hence for $n = 2$ and $n = 3$, we call (17-14) the *perturbation-theoretical integral representation* (PTIR) of $f_n(s)$. More precisely, for the self-energy, writing $\varkappa = s'$, we have

$$f_2(s) = \int_0^\infty ds' \frac{\varphi_2(s')}{s' - s} \tag{17-15}$$

a (single) spectral representation which was obtained in the axiomatic field theory [18, 19, 20, 21]. For the vertex function, we have

$$f_3(s) = \int_0^1 dz_a \int_0^1 dz_b \int_0^1 dz_c \delta(1 - z_a - z_b - z_c) \int_0^\infty d\varkappa$$
$$\times \frac{\varphi_3(z_a, z_b, z_c, \varkappa)}{\varkappa - z_a s_a - z_b s_b - z_c s_c} \tag{17-16}$$

with $g = \{a, b, c\}$.

Now, we consider the extended scattering amplitude ($n = 4$). Let $g = \{a, b, c, d\}$. Then there are seven invariant squares $s_a, s_b, s_c, s_d, s_{ab} \equiv s_{cd}, s_{ac} \equiv s_{bd}, s_{ad} \equiv s_{bc}$. Since an identity

$$s_a + s_b + s_c + s_d = s_{ab} + s_{ac} + s_{ad} \tag{17-17}$$

holds, we have to eliminate one of these seven squares. Any point $\alpha \in \Delta$ in the Feynman-parameter space belongs to one of the three regions

$$\begin{aligned} \Delta_{ad} &\equiv \{\alpha \in \Delta \mid \zeta_{ad} \leqq \zeta_{ab}, \zeta_{ad} < \zeta_{ac}\} \\ \Delta_{ab} &\equiv \{\alpha \in \Delta \mid \zeta_{ab} \leqq \zeta_{ac}, \zeta_{ab} < \zeta_{ad}\} \\ \Delta_{ac} &\equiv \{\alpha \in \Delta \mid \zeta_{ac} \leqq \zeta_{ad}, \zeta_{ac} < \zeta_{ab}\} \end{aligned} \tag{17-18}$$

For $\alpha \in \Delta_{ad}$, because of (17-17), V can be written as

$$V = \sum_{l} \alpha_l m_l^2 - \left[\sum_{\nu=a}^{d} (\zeta_\nu + \zeta_{ad})\, s_\nu + (\zeta_{ab} - \zeta_{ad})\, s_{ab} + (\zeta_{ac} - \zeta_{ad})\, s_{ac} \right] \tag{17-19}$$

In the square bracket of (17-19), the coefficients of invariant squares are positive semi-definite. In this form, we can introduce the variables $z_a, \ldots, z_{ac}$ and $\varkappa$ in the same way as in the derivation of (17-12). We can do the same thing for the other two regions Δ_{ab} and Δ_{ac}. In this way, instead of (17-14) we obtain the PTIR for the extended scattering amplitude:

$$f_4(s) = \prod_{h \neq ad} \left(\int_0^1 dz_h \right) \delta\left(1 - \sum_{h \neq ad} z_h\right) \int_0^\infty d\varkappa \frac{\varphi_{ad}(z_a, \ldots, z_{ac}, \varkappa)}{\varkappa - \sum_{h \neq ad} z_h s_h}$$

+ two other terms obtained by cyclic permutations of b, c, d (17-20)

where $h \neq ad$ means that h runs over a, b, c, d, ab, ac.

The PTIR for the scattering amplitude is obtained by putting s_a, s_b, s_c, s_d on the mass shell $s_\nu = m_\nu^2\,(\nu = a, b, c, d)$ and then by setting

$$\gamma = \left(\varkappa - \sum_\nu z_\nu m_\nu^2\right) / (z_{ab} + z_{ac})$$

$$z = z_{ab}/(z_{ab} + z_{ac}) \tag{17-21}$$

in the first term and likewise in the other terms. It is customary to write $s_{ab} \equiv s$, $s_{ac} \equiv t$, $s_{ad} \equiv u$. Then from (17-20) the scattering amplitude $f_4(s, t)$ is represented as

$$f_4(s, t) = \int_0^1 dz \int_{-\infty}^{\infty} d\gamma \frac{\varrho_{st}(z, \gamma)}{\gamma - zs - (1 - z)t}$$

$$+ \int_0^1 dz \int_{-\infty}^{\infty} d\gamma \frac{\varrho_{tu}(z, \gamma)}{\gamma - zt - (1 - z)u}$$

$$+ \int_0^1 dz \int_{-\infty}^{\infty} d\gamma \frac{\varrho_{us}(z, \gamma)}{\gamma - zu - (1 - z)s} \tag{17-22}$$

where ϱ_{st}, etc. are defined in terms of φ_{ad}, etc. If we derive (17-22) directly from (17-4), the correspondence between integration variables and the Feynman parameters are as follows:

$$\gamma \sim \left[\sum_{l} \alpha_l m_l^2 - \sum_{\nu=a}^{d} (\zeta_\nu + \zeta_{ad}) m_\nu^2 \right] / \xi$$

$$z \sim (\zeta_{ab} - \zeta_{ad})/\xi \tag{17-23}$$

with

$$\xi \equiv \zeta_{ab} + \zeta_{ac} - 2\zeta_{ad} \tag{17-24}$$

for the first term, and likewise for the other two terms.

We note that the representation (17-22) is useless if we cannot find finite quantities $\gamma_{st}(z)$, etc. such that

$$\varrho_{st}(z, \gamma) = 0 \quad \text{for } \gamma < \gamma_{st}(z), \text{ etc.} \tag{17-25}$$

because if not the Feynman integral cannot be obtained as a boundary value of $f_4(s, t)$. We see later that (17-25) can be proven in most cases of practical interest, but we should remember that it is not always true even under the stability conditions. A counterexample is provided by a complete Feynman graph Fig. 4-1(b) with the common external mass $m_\nu = m$ being larger than $\sqrt{6}$ times the common internal mass m_l [11, 22].

Finally, we consider the case $n \geqq 5$. In order to find the PTIR for $n \geqq 5$, we have to solve the following problems.

a) To find sets of $n(n-1)/2$ linearly independent squares among $2^{n-1} - 1$ ones.

b) To find the region of the Feynman-parameter space Δ in which the coefficients in $-V$ of the linearly independent squares of each set are positive semi-definite.

c) To find some combinations of the sets such that the corresponding regions found in (b) cover the whole Δ without overlapping except for their boundaries.

This program was worked out for $f_5(s)$ on the mass shell [13]. In this case, we have to find five linearly independent squares among ten. The number of such sets is found to be 162 instead of $\binom{10}{5} = 252$. By the investigation of (*b*) and (*c*), we find that it is impossible to construct PTIR in such a way that it is symmetric with respect to all external vertices. Accordingly, at least one external vertex has to be dealt with asymmetrically. If one external vertex is dealt with asymmetrically, then for any value of n we can always construct PTIR for the production amplitude in the following way [14, 23].

Let a be a particular external vertex. For $h \subset g - a$,

$$s_h = \Big(\sum_{b \in h} p_b\Big)^2 = \sum_{b \in h} p_b^2 + 2 \sum_{\{b,c\} \subset h} p_b p_c \tag{17-26}$$

We consider a set of $(n-1)(n-2)/2$ squares $s_{bc} \equiv (p_b + b_c)^2$ for $\{b, c\} \subset g - a$. Then we have

$$2p_b p_c = s_{bc} - s_b - s_c \tag{17-27}$$

Hence

$$s_h = -(n_h - 2) \sum_{b \in h} s_b + \sum_{\{b,c\} \subset h} s_{bc} \tag{17-28}$$

where n_h denotes the number of the external vertices belonging to h. For $h = g - a$, since $s_h = s_a$, from (17-28) we obtain an identity

$$\sum_{\{b,c\} \subset g-h} s_{bc} = (n-3) \sum_{b \in g-a} s_b + s_a \tag{17-29}$$

If we eliminate one of squares s_{bc} by means of (17-29), then we obtain $(n-1)(n-2)/2$ sets of $n(n-3)/2$ linearly independent squares other than $s_\nu\,(\nu \in g)$. [Note $n(n-1)/2 - n = n(n-3)/2$.] The combination of those sets satisfies the requirements stated in (b) and (c), because (17-28) leads us to

$$V = \sum_l \alpha_l m_l^2 - \sum_{b \in g-a} \eta_b s_b - \sum_{\{b,c\} \subset g-h} \xi_{bc} s_{bc} \tag{17-30}$$

where η_b is a certain linear combination of ζ_h $(h \subset g - a)$, and

$$\xi_{bc} \equiv \sum_{h \supset \{b,c\}} \zeta_h \geqq 0 \quad (h \subset g - a) \tag{17-31}$$

and because the region of Δ described in (b) is characterized by the minimum one of ξ_{bc}. In this way, we obtain

$$f_n(s) = \sum_{bc} \prod_{b'c' \neq bc} \left(\int_0^1 dz_{b'c'} \right) \delta\Big(1 - \sum_{b'c' \neq bc} z_{b'c'}\Big)$$
$$\int_{-\infty}^{\infty} d\gamma \frac{\varrho_{bc}(\{z_{b'c'} \mid b'c' \neq bc\}, \gamma)}{\gamma - \sum_{b'c' \neq bc} z_{b'c'} s_{b'c'}} \tag{17-32}$$

where any of $bc \equiv \{b, c\}$ and $b'c' \equiv \{b', c'\}$ is contained in $g - a$. For $n = 4$, (17-32) reduces to (17-22). For $n \geqq 5$, we have n different representations according to the choices of a.

17-3 Position-Space PTIR

As in Subsection 17-2, PTIR is usually considered for the momentum-space extended transition amplitude. If we regard the position-space transition amplitude as a function of invariant distances squared

$$\hat{s}_{bc} \equiv (x_b - x_c)^2 \quad \text{for } \{b, c\} \subset g \tag{17-33}$$

then from (8-33) together with (8-44), we can derive the position-space PTIR

$$\prod_{bc} \left(\int_0^1 d\hat{z}_{bc} \right) \delta\left(1 - \sum_{bc} \hat{z}_{bc}\right) \int_0^\infty d\hat{\varkappa} \frac{\hat{\varphi}_n(\{z_{bc}\}, \hat{\varkappa})}{\hat{\varkappa} - \sum_{bc} \hat{z}_{bc}\hat{s}_{bc}} \tag{17-34}$$

It is noteworthy that (17-34) consists of a single term for any value of n.

Since there are no spectral conditions in the position space, we have no restriction on the support of $\hat{\varphi}_n$. Hence we do not investigate (17-34) any more.

18-1 Analyticity

We consider the analyticity of a function defined by

$$F(s_1, \ldots, s_m) \equiv \int_{\Delta} d^{m-1}z \int_{-\infty}^{\infty} d\gamma \frac{\varphi(z_1, \ldots, z_m, \gamma)}{\gamma - \sum_{j=1}^{m} z_j s_j} \tag{18-1}$$

where

$$\Delta \equiv \{z_1, \ldots, z_m \mid z_j \geqq 0 \, (j = 1, \ldots, m), \sum_j z_j = 1\} \tag{18-2}$$

only in this section and

$$\varphi(z_1, \ldots, z_m, \gamma) = 0 \quad \text{unless } \gamma \geqq \sum_{j=1}^{m} z_j c_j \tag{18-3}$$

with some constants c_j. By linear transformations of $s_1, \ldots, s_m$ and γ, however, (18-3) can be reduced to

$$\varphi(z_1, \ldots, z_m, \gamma) = 0 \quad \text{unless } \gamma \geqq 0 \tag{18-4}$$

Hence we consider (18-1) with (18-4) without loss of generality. Then it is straightforward to see that $F(s_1, \ldots, s_m)$ is holomorphic in D, where

$$D \equiv \left\{s_1, \ldots, s_m \mid \gamma - \sum_{j=1}^{m} z_j s_j \neq 0 \text{ for all } z \in \Delta, \gamma \geqq 0\right\} \tag{18-5}$$

Let

$$\begin{aligned} D_+ &\equiv \{s_1, \ldots, s_m \mid \operatorname{Im} s_j > 0 \quad (j = 1, \ldots, m)\} \\ D_- &\equiv \{s_1, \ldots, s_m \mid \operatorname{Im} s_j < 0 \quad (j = 1, \ldots, m)\} \\ E &\equiv \{s_1, \ldots, s_m \mid \operatorname{Im} s_j = 0, \operatorname{Re} s_j < 0 \quad (j = 1, \ldots, m)\} \end{aligned} \tag{18-6}$$

Then

$$D \supset D_+ \cup D_- \cup E \tag{18-7}$$

We quote the following theorem [24] without proof.

THEOREM **18-1** Let

$$e^{i\theta} D_+ \equiv \{s_1, \ldots, s_m \mid \operatorname{Im} (e^{-i\theta} s_j) > 0 \quad (j = 1, \ldots, m)\} \tag{18-8}$$

The envelope of holomorphy (see Appendix A-2) of $D_+ \cup D_- \cup E$ is

$$D' \equiv \bigcup_{0 \leqq \theta \leqq \pi} (e^{i\theta} D_+) \tag{18-9}$$

More precisely, if two functions F_+ and F_- are holomorphic in D_+ and in D_-, respectively, and if their boundary values on E coincide with each other as distributions, then F_+ and F_- are extended into a single analytic function F holomorphic in D'.

THEOREM **18-2**

$$D' = D \tag{18-10}$$

Proof Since $\operatorname{Im} e^{-i\theta}\gamma \leqq 0$ for $0 \leqq \theta \leqq \pi$ and $\gamma \geqq 0$, we immediately have $D' \subset D$ by using the definitions (18-5) and (18-8). To show $D \subset D'$, suppose that $(s_1, \ldots, s_m) \notin D'$. Then *either* there exists $s_1 \geqq 0$, say, *or* there exist s_1 and s_2, say, such that $0 < \varphi_1 < \pi$ and $\varphi_1 + \pi < \varphi_2 < 2\pi$, where $\varphi_j \equiv \arg s_j$. For the first case, $\sum_j z_j s_j \geqq 0$ for $z_1 = 1$, $z_2 = \ldots = z_m = 0$. For the second case, $\sum_j z_j s_j \geqq 0$ for $z_1 = -c \operatorname{Im} s_2$, $z_2 = c \operatorname{Im} s_1$, $z_3 = \ldots = z_m = 0$ with $c > 0$. Thus $(s_1, \ldots, s_m) \notin D$. q.e.d.

Combining the above two theorems, we find that D is the envelope of holomorphy of $D_+ \cup D_- \cup E$. The fact that D is a natural domain of holomorphy can be proven more easily by constructing a function which is holomorphic in D but not so in an arbitrary point outside D [14].

18-2 Uniqueness

THEOREM **18-3** [14]† The integral representation (18-1) with (18-4) is unique, that is, if $F(s_1, \ldots, s_m) \equiv 0$ then $\delta(1 - \sum_j z_j)\, \varphi(z_1, \ldots, z_m, \gamma)$ vanishes as a distribution.

Proof We take the difference between the boundary values of F from D_+ and from D_-, and introduce new (real) variables $t_1 = s_1$, $t_j = s_j - s_1$ $(j = 2, \ldots, m)$. The assumption $F \equiv 0$ then yields

$$\int_{\hat{\Delta}} \prod_{j=2}^{m} dz_j \hat{\varphi}\left(z_2, \ldots, z_m, t_1 + \sum_{j=2}^{m} z_j t_j\right) \equiv 0 \tag{18-11}$$

where

$$\hat{\Delta} \equiv \left\{z_2, \ldots, z_m \mid z_j \geqq 0 \quad (j = 2, \ldots, m), \ \sum_{j=2}^{m} z_j \leqq 1\right\} \tag{18-12}$$

$$\hat{\varphi}(z_2, \ldots, z_m, \gamma) \equiv \varphi\left(1 - \sum_{j=2}^{m} z_j, z_2, \ldots, z_m, \gamma\right) \tag{18-13}$$

† The original proof was somewhat incomplete. The author thanks Prof. A. S. Wightman for pointing out this fact.

Let

$$L_j \equiv \int_{-\infty}^{t_1} dt_1 \frac{\partial}{\partial t_j} \tag{18-14}$$

where the upper limit of the integration should be put equal to t_1 after the integration is carried out. Then for any polynomial $P(z_2, \ldots, z_m)$, we have

$$P(L_2, \ldots, L_m) \int_{\hat{\Delta}} \prod_{j=2}^{m} dz_j \hat{\varphi}\left(z_2, \ldots, z_m, t_1 + \sum_{j=2}^{m} z_j t_j\right)$$

$$= \int_{\hat{\Delta}} \prod_{j=2}^{m} dz_j P(z_2, \ldots, z_m) \hat{\varphi}\left(z_2, \ldots, z_m, t_1 + \sum_{j=2}^{m} z_j t_j\right) \tag{18-15}$$

Hence setting $t_1 = \gamma$ and $t_2 = \ldots = t_m = 0$, we obtain

$$\int_{\hat{\Delta}} \prod_{j=2}^{m} dz_j \, P(z_2, \ldots, z_m) \hat{\varphi}\,(z_2, \ldots, z_m, \gamma) = 0 \tag{18-16}$$

Since $\hat{\varphi}$ is a distribution whose support is compact with respect to $z_2, \ldots, z_m$, there exists a function $\psi(z_2, \ldots, z_m, \gamma)$ continuous with respect to $z_2, \ldots, z_m$ such that its certain derivative equals $\hat{\varphi}$ (see Appendix A-1). Since any derivative of any polynomial is a polynomial, (18-16) implies

$$\int_{\hat{\Delta}} \prod_{j=2}^{m} dz_j \, \tilde{P}(z_2, \ldots, z_m) \psi(z_2, \ldots, z_m, \gamma) = 0 \tag{18-17}$$

for an arbitrary polynominal $\tilde{P}(z_2, \ldots, z_m)$.

To show $\hat{\varphi} = 0$ is equivalent to proving

$$\int_{-\infty}^{\infty} d\gamma \, g(\gamma) \int_{\hat{\Delta}} \prod_{j=2}^{m} dz_j \cdot f(z_2, \ldots, z_m) \hat{\varphi}(z_2, \ldots, z_m, \gamma) = 0 \tag{18-18}$$

for arbitrary test functions $g(\gamma)$ and $f(z_2, \ldots, z_m)$. By integrating by parts, (18-18) is rewritten as

$$\int_{-\infty}^{\infty} d\gamma \, g(\gamma) \int_{\hat{\Delta}} \prod_{j=2}^{m} dz_j \cdot \tilde{f}(z_2, \ldots, z_m) \psi(z_2, \ldots, z_m, \gamma) = 0 \tag{18-19}$$

where $\tilde{f}$ is a certain derivative of f. According to Weierstrass' approximation theorem, given $\varepsilon > 0$, we can always find a polynomial $\tilde{P}$ such that

$$\tilde{f}(z_2, \ldots, z_m) = \tilde{P}(z_2, \ldots, z_m) + \varepsilon \tilde{Q}(z_2, \ldots, z_m) \tag{18-20}$$

with $|\tilde{Q}| \leqq 1$ in the compact region $\hat{\Delta}$. Then we have

$$\int_{-\infty}^{\infty} d\gamma\, g(\gamma) \int_{\hat{\Delta}} \prod_{j=2}^{m} dz_j \tilde{f}(z_2, \ldots, z_m)\, \psi(z_2, \ldots, z_m, \gamma)$$

$$= \varepsilon \int_{\hat{\Delta}} \prod_{j=2}^{m} dz_j \tilde{Q}(z_2, \ldots, z_m) \int_{-\infty}^{\infty} d\gamma\, g(\gamma)\, \psi(z_2, \ldots, z_m, \gamma) \tag{18-21}$$

because of (18-17). Since the absolute value of the right-hand side of (18-21) becomes arbitrarily small, we obtain (18-19). Thus $\hat{\varphi} = 0$, namely, $\delta\left(1 - \sum_j z_j\right)$ $\varphi = 0$. q.e.d.

The above theorem assures the uniqueness of the PTIR consisting of a single term. When PTIR consists of several terms, we cannot in general prove its uniqueness, but we have the following theorem for $n = 4$.

THEOREM **18-4** [14] In (17-22), if

$$\begin{aligned} \varrho_{st}(z, \gamma) &= 0 \quad \text{unless} \quad \gamma \geqq zc_s + (1 - z)\, c_t \\ \varrho_{tu}(z, \gamma) &= 0 \quad \text{unless} \quad \gamma \geqq zc_t + (1 - z)\, c_u \\ \varrho_{us}(z, \gamma) &= 0 \quad \text{unless} \quad \gamma \geqq zc_u + (1 - z)\, c_s \end{aligned} \tag{18-22}$$

where c_s, c_t, c_u are certain constants satisfying

$$c_s + c_t + c_u > c_0 \equiv \sum_{\nu=a}^{d} m_\nu^2 \tag{18-23}$$

then $f_4(s, t)$ uniquely determines the weight distributions ϱ_{st}, ϱ_{tu}, and ϱ_{us} except for the contributions from $z = 0$ and $z = 1$.

Proof Let f_{st}, f_{tu}, and f_{us} be the first, the second, and the third term of (17-22), respectively, which are holomorphic in D_{st}, in D_{tu}, and in D_{us}, respectively. Here D_{st} is the complement of

$$D_{st}^* \equiv \{s, t \mid z(s - c_s) + (1 - z)(t - c_t) \geqq 0 \quad \text{for some} \quad 0 \leqq z \leqq 1\} \tag{18-24}$$

and D_{tu} and D_{us} are similar. If $f_4 \equiv 0$ then $f_{st} = -f_{tu} - f_{us}$ is holomorphic in

$$\hat{D} \equiv D_{st} \cup (D_{tu} \cap D_{us}) \tag{18-25}$$

With the aid of (18-23), we can show after some manipulation [15] that $\hat{D} = D_s \cap D_t$, where D_s is the complement of

$$D_s^* \equiv \{s, t \mid \operatorname{Im} s = 0, \quad \operatorname{Re} s \geqq c_s\} \tag{18-26}$$

and D_t and D_u are similar. Likewise, f_{tu} and f_{us} are holomorphic in $D_t \cap D_u$ and in $D_u \cap D_s$, respectively. Since the sum of the three functions vanishes, it is easy to show, with the aid of (18-23) again, that

$$f_{st} = f_s - f_t, \; f_{tu} = f_t - f_u, \; f_{us} = f_u - f_s \tag{18-27}$$

where f_s, f_t, f_u are function holomorphic in D_s, in D_t, in D_u, respectively. Hence, Theorem 18-3 implies that

$$\varrho_{st}(z, \gamma) = \varrho_{tu}(z, \gamma) = \varrho_{us}(z, \gamma) = 0 \quad \text{for} \quad 0 < z < 1 \tag{18-28}$$

q.e.d.

In the above theorem, the condition (18-23) is essential. If it holds, a real region

$$D_0 \equiv \{s, t \mid \operatorname{Im} s = \operatorname{Im} t = 0, \operatorname{Re} s < c_s, \operatorname{Re} t < c_t, \operatorname{Re} u < c_u\} \tag{18-29}$$

is non-empty. This region is called the *Symanzik region* [25]. According to (17-22), $f_4(s, t)$ is holomorphic in the Symanzik region D_0.

It is interesting to compare (17-22) with the *Mandelstam representation* [5], which reads

$$\begin{aligned} f_4(s, t) = & \int_{c_s}^{\infty} ds' \int_{c_t}^{\infty} dt' \frac{\sigma_{st}(s', t')}{(s' - s)(t' - t)} + \int_{c_t}^{\infty} dt' \int_{c_u}^{\infty} du' \frac{\sigma_{tu}(t', u')}{(t' - t)(u' - u)} \\ & + \int_{c_u}^{\infty} du' \int_{c_s}^{\infty} ds' \frac{\sigma_{us}(u', s')}{(u' - u)(s' - s)} + \int_{c_s}^{\infty} ds' \frac{\sigma_s(s')}{s' - s} \\ & + \int_{c_t}^{\infty} dt' \frac{\sigma_t(t')}{t' - t} + \int_{c_u}^{\infty} du' \frac{\sigma_u(u')}{u' - u} \end{aligned} \tag{18-30}$$

in the unsubtracted form, where σ_{st}, σ_{tu}, and σ_{us} are called *double spectral functions*. Evidently, if (18-30) holds, $f_4(s, t)$ is holomorphic in $D_s \cap D_t \cap D_u$, which is much larger than $D_{st} \cap D_{tu} \cap D_{us}$. We can convert (18-30) into the form of (17-22) by means of the Feynman identity. We obtain

$$\varrho_{st}(z, \gamma) = \int_{c_s}^{\infty} ds' \int_{c_t}^{\infty} dt' \sigma_{st}(s,' t') \, \delta'(\gamma - zs' - (1 - z) t') + \delta(z)\sigma_t(\gamma), \quad \text{etc.} \tag{18-31}$$

According to Theorem 18-4, the relations (18-31) are unique if we assume that ϱ_{st}, etc. have no δ-functions at $z = 1$.

If $f_4(s, t)$ given by (17-22) has some symmetry property (crossing symmetry), we can express it uniquely in terms of the weight distributions. For example, if $f_4(s, t) = f_4(t, s)$ then Theorem 18-4 implies that

$$\begin{aligned} \varrho_{st}(z, \gamma) &= \varrho_{st}(1 - z, \gamma) \\ \varrho_{tu}(z, \gamma) &= \varrho_{us}(1 - z, \gamma) \\ \varrho_{us}(z, \gamma) &= \varrho_{tu}(1 - z, \gamma) \end{aligned} \tag{18-32}$$

for $0 < z < 1$.

18-3 Other Properties

In (18-1), when $\varphi(z_1, \ldots, z_m, \gamma)$ does not decrease sufficiently as $\gamma \to \infty$, *subtraction procedure* is necessary to obtain a convergent integral. The subtracted form of (18-1) is

$$F(s_1, \ldots, s_m) = P(s_1, \ldots, s_m) + \int_\Delta d^{m-1}z \int_{-\infty}^{\infty} d\gamma \left[\frac{\sum_{j=1}^{m} (s_j - a_j)}{\gamma - \sum_{j=1}^{m} z_j a_j} \right]^k \frac{\varphi(z_1, \ldots, z_m, \gamma)}{\gamma - \sum_{j=1}^{m} z_j s_j} \tag{18-33}$$

where k is a positive integer, $P(s_1, \ldots, s_m)$ is a polynomial of degree $k - 1$, and a point $(a_1, \ldots, a_m)$ lies in D defined by (18-5). When (18-1) converges, the difference between (18-1) and (18-33) is a polynomial (in contrast with the case of the Mandelstam representation).

It is quite an interesting property that the unsubtracted PTIR can describe a function $F(s_1, \ldots, s_m)$ which increases in particular directions [16]. For example, if

$$\varphi(z_1, \ldots, z_m, \gamma) = Y_\nu(z_1)\, \tilde{\varphi}(z_2, \ldots, z_m, \gamma) \tag{18-34}$$

(see Appendix A-1), then for $\operatorname{Re} \nu < 1$ (18-1) behaves like $s_1^{-\nu}$ as $|s_1| \to \infty$. Roughly speaking, subtraction procedure is necessary for the functions which are non-decreasing in *all* directions (in D).

Given a function F which is holomorphic in D and has certain asymptotic behavior, we can find the weight distribution φ in terms of F [16]. The extension of this problem to the case (17-22) is solved partially [15].

The PTIR is useful mainly for the quantities off the mass shell, e.g., for the analysis of the Bethe–Salpeter equation. For the amplitude on the mass shell, (17-22) is useful only as a theoretical tool, e.g., in the proof of the dispersion relations. This is because PTIR does not exhibit the analyticity in a complex neighborhood of the physical region. Therefore it is difficult to combine (17-22) with the unitarity of the S-matrix, in contrast with the Mandelstam representation.

§ 19 SUPPORT PROPERTIES IN THE GENERAL-MASS CASE

The most important problem in the theory of PTIR is to find the supports of the weight distributions. This and the succeeding three sections are devoted to the investigation of support properties of PTIR.

19-1 General Method

Given a Feynman graph G, the corresponding Feynman function $f_G(s)$ can be rewritten in the form of PTIR, and therefore the support of each weight distribution, say, $\bar{\varphi}_G(z, \varkappa)$, of it is well defined. The problem which we consider in this section is to investigate

$$\text{Supp}\,[\Phi_n] \equiv \bigcup_{G \in \Phi_n} \text{supp}\,\bar{\varphi}_G \tag{19-1}$$

It is very difficult in most cases to find (19-1) exactly, and we have to content ourselves by finding an upper bound on it.

An upper bound on supp $\bar{\varphi}_G$ can be found in the following way. First, we find a point $p = \bar{p}$ in the complex $4(n-1)$-dimensional momentum space such that

$$V(\bar{p}, \alpha) \geqq 0 \quad \text{for all} \quad \alpha \in \Delta \tag{19-2}$$

Second, as done in Subsection 17-2, we rewrite V as

$$V(p, \alpha) = \sum_l \alpha_l m_l^2 - \sum_{j=1}^{n(n-1)/2} \eta_j(\alpha)\, s_j \tag{19-3}$$

where $\eta_j(\alpha) \geqq 0$ for the relevant values of α and $s_1, \ldots, s_{n(n-1)/2}$ are independent squares of p. Let c_j be the value of s_j at $p = \bar{p}$. Then (19-2) implies

$$\sum_l \alpha_l m_l^2 \geqq \sum_j \eta_j c_j \tag{19-4}$$

Since $\bar{\varphi}_G(z, \varkappa)$ has contributions only from $\varkappa = \sum_l \alpha_l m_l^2/\xi$ and $z_j = \eta_j/\xi$ $(j = 1, \ldots, n(n-1)/2)$, where $\xi \equiv \sum_j \eta_j$, (19-4) yields that

$$\bar{\varphi}_G(z, \varkappa) = 0 \quad \text{unless} \quad \varkappa \geqq \sum_j z_j c_j \tag{19-5}$$

Thus it is important to find a point $p = \bar{p}$ such that (19-2) holds for *all* $G \in \Phi_n$. For this purpose, the following theorem is useful.

THEOREM **19-1** [26] If the set of external momenta $\bar{p}$ is euclidean (see Definition 13-5 and also Theorem 13-6), and if we can find internal momenta q_l ($l \in |G|$) such that

(a) they are linear combinations of $\bar{p}$ and satisfy the conservation law (12-4) at each vertex, and

(b) $m_l^2 \geqq q_l^2$ for all $l \in |G|$,
then (19-2) holds.

Proof Since $\{q_l \mid l \in |G|\}$ is euclidean, the circuit representation (7-26) of V implies

$$V(\bar{p}, \alpha) \geqq \sum_l \alpha_l(m_l^2 - q_l^2) \geqq 0 \tag{19-6}$$

q.e.d.

In particular, if all momenta $\bar{p}_b$ are timelike and parallel, each q_l can be regarded as a one-dimensional quantity. In this case, therefore, if we regard m_l as a capacity $c(l)$, $\bar{p}_b$ as a demand $d(b)$, and q_l as a flow $f(l)$, the problem of finding $\{q_l \mid l \in |G|\}$ is exactly reduced to the transport problem discussed in Section 5. Accordingly, Theorem 5-3 is translated as follows.

THEOREM **19-2** When all external momenta $\bar{p}_b$ are timelike and parallel, a necessary and sufficient condition for the existence of q_l ($l \in |G|$) satisfying the conditions of Theorem 19-1 is

$$\Big(\sum_{b\in h} \bar{p}_b\Big)^2 \leqq \Big(\sum_{l\in S} m_l\Big)^2$$

$$\text{for any} \quad h \subset g \quad \text{and any} \quad S \in \mathcal{S}(h \mid g - h) \tag{19-7}$$

If we combine Theorem 19-2 with the spectral conditions (17-1), we obtain the following theorem.

THEOREM **19-3** For $\bar{p}$ timelike and parallel, (19-2) holds for all $G \in \Phi_n$ if

$$\Big(\sum_{b\in h} \bar{p}_b\Big)^2 \leqq M_h^2 \quad \text{for any} \quad h \subset g \tag{19-8}$$

For $n = 2$, let a and b be the two external vertices. From Theorems 19-1 and 19-2 together with (7-45), we have

$$\sum_l \alpha_l m_l^2 \geqq (W^{(a|b)}/U) \min_{S\in\mathcal{S}(a|b)} \Big(\sum_{l\in S} m_l\Big)^2 \tag{19-9}$$

In particular, if all internal masses are equal to each other, (19-9) reduces to

$$W^{(a|b)}/U \leqq 1/r^2 \quad \text{for} \quad \alpha \in \Delta \tag{19-10}$$

where

$$r \equiv \min_{S\in\mathcal{S}(a|b)} N(S) \tag{19-11}$$

If we apply Theorem 19-3 to the case $n = 2$, we obtain

$$\varphi_2(s') = 0 \quad \text{unless} \quad s' \geqq M_a^2 \tag{19-12}$$

Since $\varphi_G(s') \neq 0$ for $s' > M_a^2$ in the $N = 2$ self-energy graph G, Supp $[\Phi_2]$ is given by $s' \geqq M_a^2$. This result is well known in the axiomatic field theory [19, 20, 21].

19-2 Vertex function

In this subsection, we discuss the support properties of $\varphi_3(z_a, z_b, z_c, \varkappa)$ involved in (17-16) by applying Theorem 19-3 to Φ_3.

Since $\bar{p}_a, \bar{p}_b, \bar{p}_c$ are parallel, we can write

$$\bar{p}_\nu = \lambda_\nu \hat{p} \quad (\nu = a, b, c) \tag{19-13}$$

where $\hat{p}^2 = 1$ and $\lambda_a, \lambda_b, \lambda_c$ are constants satisfying

$$\lambda_a + \lambda_b + \lambda_c = 0 \tag{19-14}$$

The conditions (19-8) are written as

$$|\lambda_\nu| \leqq M_\nu \quad (\nu = a, b, c) \tag{19-15}$$

With those values of λ_ν, Theorem 19-3 implies

$$\varphi_3(z_a, z_b, z_c, \varkappa) = 0 \quad \text{unless} \quad \varkappa \geqq \sum_\nu z_\nu \lambda_\nu^2 \tag{19-16}$$

Thus our task is to maximize $\sum_\nu z_\nu \lambda_\nu^2$ under the restrictions (19-14) and (19-15). For simplicity, we assume that M_a, M_b, M_c satisfy the triangular inequalities

$$|M_a - M_b| \leqq M_c \leqq M_a + M_b \tag{19-17}$$

use of which will be justified in the next subsection.

By means of Lagrange multiplier, we find the extremum condition yields $\sum_\nu z_\nu^{-1} = 0$ for $\prod_\nu z_\nu \neq 0$, that is, it contradicts $z_\nu > 0$ for $\nu = a, b, c$; the exceptional case $\prod_\nu z_\nu = 0$ implies $\sum_\nu z_\nu \lambda_\nu^2 = 0$. Therefore, the maximum of $\sum_\nu z_\nu \lambda_\nu^2$ has to occur at one of the three extreme points of the region defined by (19-14) and (19-15). Thus $\varphi_3 = 0$ unless

$$\begin{aligned} \varkappa \geqq \max [&z_a M_a^2 + z_b M_b^2 + z_c (M_a - M_b)^2, \\ &z_a M_a^2 + z_b (M_a - M_c)^2 + z_c M_c^2, \\ &z_a (M_b - M_c)^2 + z_b M_b^2 + z_c M_c^2] \end{aligned} \tag{19-18}$$

where "max" should be taken for each (z_a, z_b, z_c) *fixed*; hereafter we always use max and min in this sense if they appear in the expressions for support properties. Unfortunately, (19-18) is not exactly Supp $[\Phi_3]$, but (19-18) is a good upper bound on it.

For $s_c = m_c^2$, if it satisfies the stability condition $m_c \leqq M_c$, setting

$$z = \frac{z_a}{z_a + z_b}, \qquad \gamma = \frac{\varkappa - z_c m_c^2}{z_a + z_b} \tag{19-19}$$

in (17-16), we obtain a two-variable representation

$$f_3(s_a, s_b, m_c^2) = \int_0^1 dz \int_0^\infty d\gamma \frac{\varphi_{ab}(z, \gamma)}{\gamma - z s_a - (1 - z)\, s_b} \tag{19-20}$$

where

$$\varphi_{ab}(z, \gamma) \equiv \int_0^1 dy\, y\, \varphi_3(yz, y(1 - z), 1 - y, y\gamma + (1 - y)\, m_c^2) \tag{19-21}$$

and it vanishes unless

$$\gamma \geqq \max\,[z M_a^2 + (1 - z)\,(M_a - M_c)^2,$$
$$z(M_b - M_c)^2 + (1 - z)\, M_b^2] \tag{19-22}$$

In (19-22), under the stability condition, M_c can be replaced by its smallest possible value m_c if $m_c \geqq |M_a - M_b|$. Then, in contrast with the three-variable case, (19-22) is realized in the triangle graph (see Fig. 13-4). In this example, (19-20) is directly obtained from the Feynman-parametric formula by a transformation of variables

$$z = \frac{\alpha_2}{\alpha_1 + \alpha_2}, \qquad \gamma = \frac{\sum_j \alpha_j m_j^2 - \alpha_1 \alpha_2 m_c^2}{(\alpha_1 + \alpha_2)\, \alpha_3} \tag{19-23}$$

Since $\alpha_1 + \alpha_2 + \alpha_3 = 1$,

$$\gamma = \frac{[\varrho(z)]^2}{\alpha_3} + \frac{m_3^2}{1 - \alpha_3} + z(1 - z)\, m_c^2 \tag{19-24}$$

where

$$\varrho(z) \equiv (1 - z)\, m_1^2 + z m_2^2 - z(1 - z)\, m_c^2 \geqq 0 \tag{19-25}$$

Hence

$$\inf_{\alpha_3} \gamma = [\varrho(z) + m_3]^2 + z(1 - z)\, m_c^2 \tag{19-26}$$

If $m_c = M_c = m_1 + m_2$ then

$$\inf_{\alpha_3} \gamma = (1 - z) m_1^2 + z m_2^2 + m_3^2 + 2m_3 \, |m_1(1 - z) - m_2 z|$$
$$= \max [z(m_2 + m_3)^2 + (1 - z)(m_1 - m_3)^2, \quad z(m_2 - m_3)^2 + (1 - z)(m_1 + m_3)^2] \qquad (19\text{-}27)$$

Since $M_a = m_2 + m_3$, $M_b = m_1 + m_3$, and $M_c = m_1 + m_2$, (19-27) coincides with (19-22).

It is interesting to rewrite (19-20) together with (19-22) and $M_c = m_c$ into another form. Let $P = p_c$ and $p \equiv (p_b - p_a)/2$, so that

$$s_a = (\tfrac{1}{2}P + p)^2, \quad s_b = (\tfrac{1}{2}P - p)^2, \quad m_c^2 = \mathrm{P}^2 \qquad (19\text{-}28)$$

By a transformation of variables

$$z = (1 + \zeta)/2, \quad \gamma = \beta + \tfrac{1}{4}(1 - \zeta^2)\, m_c^2 \qquad (19\text{-}29)$$

we have

$$f_3(P, p) = \int_{-1}^{1} d\zeta \int_{0}^{\infty} d\beta \, \frac{\tilde{\varphi}_{ab}(\zeta, \beta)}{\beta - (p + \tfrac{1}{2}\zeta P)^2} \qquad (19\text{-}30)$$

where $\tilde{\varphi}_{ab}(\zeta, \beta) = 0$ unless

$$\beta \geqq \max \{[M_a - \tfrac{1}{2}(1 - \zeta)\, m_c]^2, \quad [M_b - \tfrac{1}{2}(1 + \zeta)\, m_c]^2\} \qquad (19\text{-}31)$$

The integral representation (19-30) together with (19-31) is the one proposed (but not correctly proven) in the axiomatic field theory [7, 8, 9]. Thus it is true in every order of perturbation theory [10]. If m_c does not satisfy the stability condition, however, this representation is no longer true even in the triangle graph (see Appendix B-2).

19-3 Extended Scattering Amplitude

In this subsection, we discuss the support properties of φ_{ad} involved in (17-20); those of φ_{ab} and φ_{ac} are similar.

As before, we set $\bar{p}_\nu = \lambda_\nu \hat{p}\ (\nu = a, b, c, d)$, where

$$\lambda_a + \lambda_b + \lambda_c + \lambda_d = 0 \qquad (19\text{-}32)$$

and $\hat{p}^2 = 1$. Then (19-8) reads

$$|\lambda_\nu| \leqq M_\nu \quad (\nu = a, b, c, d)$$
$$|\lambda_a + \lambda_\nu| \leqq M_{a\nu} \quad (\nu = b, c, d) \qquad (19\text{-}33)$$

Theorem 19-3 then implies that $\varphi_{ad}(z_a, \ldots, z_{ac}, \varkappa) = 0$ unless

$$\varkappa \geqq \sum_{v=a}^{d} z_v \lambda_v^2 + z_{ab}(\lambda_a + \lambda_b)^2 + z_{ac}(\lambda_a + \lambda_c)^2 \tag{19-34}$$

Since the extremum condition for the right-hand side of (19-34) under (19-32) yields a particular relation for $z_a, \ldots, z_{ac}$, it is probably unimportant. Hence we consider the extreme points of the region defined by (19-32) and (19-33) as before. For example, if we put

$$\lambda_a = M_a, \quad \lambda_b = -M_b, \quad \lambda_a + \lambda_c = M_{ac} \tag{19-35}$$

(19-34) reduces to

$$\varkappa \geqq z_a M_a^2 + z_b M_b^2 + z_c(M_{ac} - M_a)^2 + z_d(M_b - M_{ac})^2$$
$$+ z_{ab}(M_a - M_b)^2 + z_{ac} M_{ac}^2 \tag{19-36}$$

However, the point (19-35) does not necessarily satisfy all of the inequalities (19-33), that is, (19-36) is true *under the additional assumptions*

$$|M_{ac} - M_a| \leqq M_c, \quad |M_b - M_{ac}| \leqq M_d$$
$$|M_a - M_b| \leqq M_{ab}, \quad M_c - M_b \leqq M_{ad} \equiv M_{bc} \tag{19-37}$$

and

$$M_a + M_b \leqq M_{ac} + M_{ad} \tag{19-38}$$

As shown below (see Theorem 19-8), the triangular inequalities (19-37) are always satisfied for any reasonable set Φ_n, but (19-38) is a very stringent assumption on Φ_4 as is seen by the following example.

Example Consider the set of all Feynman graphs G having four external lines of mass m such that all vertices of G have F-degree 3 and the lines incident with each vertex have masses m, m, 1, and suppose $m > 1$. [This is a model corresponding to the nucleon-nucleon scattering.] Then we have $M_a = M_b = m + 1$ and $M_{ac} = M_{ad} = 2$, and therefore (19-38) is not satisfied. It should be noticed, however, that both $M_{ac} = 2$ and $M_{ad} = 2$ are never realized in a *single* Feynman graph.

As is suggested by the above example, it is better to prove the inequality corresponding to (19-38) in *each* Feynman graph and later consider the spectral conditions of Φ_n. Since originally the spectral conditions are determined implicitly from a certain interaction Lagrangian, it is quite reasonable to suppose that Φ_n satisfies the following condition.

ASSUMPTION **19-4** Let $g = \{a_1, \ldots, a_n\}$. For any $G_1, \ldots, G_m \in \Phi_n$, we require $G \in \Phi_n$, where G is any connected Feynman graph having n external vertices $a_1, \ldots, a_n$ such that it can be constructed out of $G_1, \ldots, G_m$ through repeated use of the following operation: to replace two external lines associated with a_j belonging to two graphs by an internal line† incident with the relevant two vertices (in short, to "join" two external lines of the same kind).

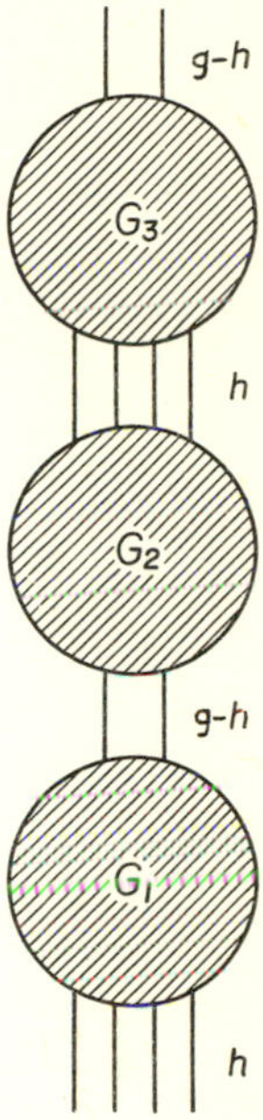

Fig. 19-1 A Feynman graph composed of G_1, G_2, G_3.

For example, if $G_1, G_2, G_3 \in \Phi_n$ then the above assumption implies that the Feynman graphs shown in Fig. 19-1 and in Fig. 19-2 belong to Φ_n.

Given a Feynman graph G, we introduce a complete Feynman graph‡, $\bar{G}$, in the generalized sense (that is, we admit parallel lines, loop lines, and internal vertices) in such a way that

$$\bar{G} - I_0 = G \quad \text{with} \quad I_0 \equiv \{l \mid m_l = 0\} \tag{19-39}$$

and define

$$\bar{\Phi}_n \equiv \{\bar{G} \mid G \subset \Phi_n\} \tag{19-40}$$

† We need not specify its mass.

‡ Any cut-set in $\bar{G}$ corresponds to a cut in G.

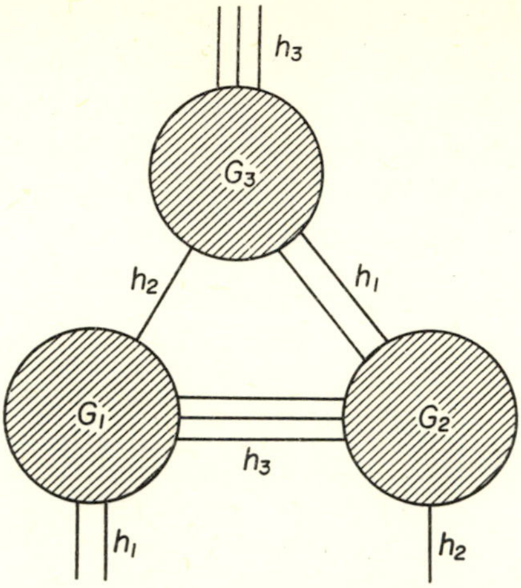

Fig. 19-2 Another Feynman graph composed of G_1, G_2, G_3, where $h_1 \cup h_2 \cup h_3 = g$.

Furthermore, we introduce the following notation:

$$m(I) \equiv \sum_{l \in I} m_l \tag{19-41}$$

$$M_h(G) = M_{g-h}(G) \equiv \min_{S \in S_G(h|g-h)} m(S) \tag{19-42}$$

where $M_h(G) = \infty$ if $S_G(h \mid g - h) = \emptyset$, and $M_h(\bar{G})$ is similar. For any $S \in S_G(h \mid g - h)$, we can always find $\bar{S} \in S_{\bar{G}}(h \mid g - h)$ such that $\bar{S} \subset S \cup I_0$. Since $m(S) = m(\bar{S})$ because $m(I_0) = 0$, we have

$$M_h(G) \geqq M_h(\bar{G}) \geqq 0 \tag{19-43}$$

The spectral conditions (17-1) imply

$$M_h = \inf_{G \in \Phi_n} M_h(G) \tag{19-44}$$

THEOREM **19-5** [11]

$$M_h = \inf_{\bar{G} \in \bar{\Phi}_n} M_h(\bar{G}) \tag{19-45}$$

Proof Because of (19-43), it is sufficient to show that for any $G \in \Phi_n$ and for any $h \subset g$ there exists $G' \in \Phi_n$ such that

$$M_h(\bar{G}) \geqq M_h(G') \tag{19-46}$$

Such a Feynman graph G' is given by Fig. 19-1 with $G_1 = G_2 = G_3 = G$. Indeed, for any $\bar{S} \in \boldsymbol{S}_{\bar{G}}(h \mid g - h)$, by regarding $\bar{S}$ as a cut-set in $\bar{G}_2$, we can find a cut-set S' in G' such that $S' \subset \bar{S}$ [hence $m(S') \leqq m(\bar{S})$] because of Theorem 2-1(d) ($\bar{S} \not\subset I_0$ because G is connected), and moreover $S' \in \boldsymbol{S}_{G'}(h \mid g - h)$ because S' is a cut-set in $G'/(|G_1| + |G_3|)$ [see Theorem 2-2(d)]. q.e.d.

THEOREM **19-6** For a complete graph $\bar{G}$, if $\bar{S} \in \boldsymbol{S}_{\bar{G}}(h \mid g - h)$ and $\bar{S}' \in \boldsymbol{S}_{\bar{G}}(h' \mid g - h')$ for any $h, h' \subset g$, then we have

$$\bar{S} \oplus \bar{S}' \in \boldsymbol{S}_{\bar{G}}[h \oplus h' \mid g - (h \oplus h')] \tag{19-47}$$

where the symbol $\oplus$ denotes the symmetric difference (see Subsection 1-1).

Proof We consider a reduced graph $R \equiv \bar{G}/(|\bar{G}| - \bar{S} \cup \bar{S}')$, which is also a complete graph and has only four vertices; they may naturally be called $h \cap h'$, $h \cap (g - h')$, $(g - h) \cap h'$, and $(g - h) \cap (g - h')$. Since R is essentially the graph shown in Fig. 4-1(b), it is easy to see that $\bar{S} \oplus \bar{S}'$ is a cut-set belonging to $\boldsymbol{S}_R(h \cap h',\ (g - h) \cap (g - h') \mid h \cap (g - h'),\ (g - h) \cap h')$. Hence in $\bar{G}$, $\bar{S} \oplus \bar{S}'$ is a cut-set because of Theorem 2-2(d) and belongs to $\boldsymbol{S}_{\bar{G}}[h \oplus h' \mid g - (h \oplus h')]$ because the division $[h \oplus h' \mid g - (h \oplus h')]$ is nothing but the division $([h \cap (g - h')] \cup [(g - h) \cap h'] \mid [h \cap h'] \cup [(g - h) \cap (g - h')])$. q.e.d.

THEOREM **19-7** For any $h, h' \subset g$,

$$M_h(\bar{G}) + M_{h'}(\bar{G}) \geqq M_{h\oplus h'}(\bar{G}) \geqq |M_h(\bar{G}) - M_{h'}(\bar{G})| \tag{19-48}$$

Proof

$$\begin{aligned} M_h(\bar{G}) + M_{h'}(\bar{G}) &= \min_{\substack{\bar{S} \in \boldsymbol{S}_{\bar{G}}(h \mid g-h) \\ \bar{S}' \in \boldsymbol{S}_{\bar{G}}(h' \mid g-h')}} [m(\bar{S}) + m(\bar{S}')] \\ &\geqq \min m(\bar{S} \oplus \bar{S}') \\ &\geqq \min_{\bar{S} \in \boldsymbol{S}_{\bar{G}}[h\oplus h' \mid g-(h\oplus h')]} m(\bar{S}) \\ &= M_{h\oplus h'}(\bar{G}) \end{aligned} \tag{19-49}$$

with the aid of Theorem 19-6. The latter inequality of (19-48) follows from the former one by substituting $h \oplus h'$ for h or h'. q.e.d.

The corresponding formula for $M_h(G)$ cannot be proven because for $S, S' \in \boldsymbol{S}$, $S \oplus S'$ is not necessarily a cut-set in G.

THEOREM **19-8** If $h \cap h' = \emptyset$ for $h, h' \subset g$, then

$$M_h + M_{h'} \geqq M_{h \cup h'} \geqq |M_h - M_{h'}| \tag{19-50}$$

Proof According to (19-45), for any $\varepsilon > 0$ we have $\bar{G}_1, \bar{G}_2, \bar{G}_3, \in \bar{\Phi}_n$ such that

$$\begin{gathered} |M_h - M_h(\bar{G}_1)| < \varepsilon, \quad |M_{h'} - M_{h'}(\bar{G}_2)| < \varepsilon \\ |M_{h \cup h'} - M_{h \cup h'}(\bar{G}_3)| < \varepsilon \end{gathered} \tag{19-51}$$

We construct $\bar{G} \in \bar{\Phi}_n$ from $\bar{G}_1, \bar{G}_2, \bar{G}_3$ as in Fig. 19-2† with $h_1 = h$, $h_2 = h'$, $h_3 = g - h \cup h'$. Then $\bar{G} \in \bar{\Phi}_n$ and

$$\begin{gathered} |M_h - M_h(\bar{G})| < \varepsilon, \ |M_{h'} - M_{h'}(\bar{G})| < \varepsilon \\ |M_{h \cup h'} - M_{h \cup h'}(\bar{G})| < \varepsilon \end{gathered} \tag{19-52}$$

Since ε can be arbitrarily small, Theorem 19-7 implies (19-50) because $h \oplus h' = h \cup h'$ for $h \cap h' = \emptyset$. q.e.d.

In particular, for any $a, b \in g$ we obtain the triangular inequalities

$$M_a + M_b \geqq M_{ab} \geqq |M_a - M_b| \tag{19-53}$$

No proof of (19-53) seems to exist (but it was used tacitly [27]) in the axiomatic field theory. We also note that if $h \cap h' \neq \emptyset$ then $M_h + M_{h'} \geqq M_{h \oplus h'}$ does not hold in general.

Now, we return to the case $n = 4$ $(g = \{a, b, c, d\})$.

THEOREM **19-9** [11]

$$\sum_{v=a}^{d} M_v(\bar{G}) \leqq M_{ab}(\bar{G}) + M_{ac}(\bar{G}) + M_{ad}(\bar{G}) \tag{19-54}$$

Proof We may assume without loss of generality that a, b, c, d are distinct because otherwise the right-hand side is infinite. We have only to show that for any $\bar{S}_h \in S_{\bar{G}}(h \mid g - h)$ $(h = ab, ac, ad)$ there exist $S_v \in S_{\bar{G}}(v \mid g - v)$ $(v = a, b, c, d)$ such that

$$\sum_{v=a}^{d} N(\bar{S}_v \cap I) \leqq N(\bar{S}_{ab} \cap I) + N(\bar{S}_{ac} \cap I) + N(\bar{S}_{ad} \cap I) \tag{19-55}$$

with $I \equiv \{l\}$ for any $l \in |\bar{G}|$. We define $\bar{S}_v$ $(v = a, b, c, d)$ by the stars $S[v]$ in a reduce graph $R \equiv \bar{G}/(|\bar{G}| - \bar{S}_{ab} \cup \bar{S}_{ac} \cup \bar{S}_{ad})$. Then because of Theorem

† Use of this graph was suggested to the author by Prof. H. Araki.

2-2(d), $\bar{S}_\nu \in S_{\bar{G}}(\nu \mid g - \nu)$. Let $N_1(l)$ and $N_2(l)$ be the left-hand and the right-hand side of (19-55), respectively. It is evident that $N_1(l) \leqq 2$, because a, b, c, d are distinct in R. If $N_1(l) \geqq 1$ then

$$l \in \bigcup_{\nu=a}^{d} \bar{S}_\nu \subset \bar{S}_{ab} \cup \bar{S}_{ac} \cup \bar{S}_{ad} \tag{19-56}$$

whence $N_2(l) \geqq 1$. Finally, if $N_1(l) = 2$ then $l \in \bar{S}_a \cap \bar{S}_b$, say, that is, $l \in S[a] \cap S[b]$ in R. That means that a and b are adjacent in R, and therefore $l \in \bar{S}_{ac}$ and $l \in \bar{S}_{ad}$, that is, $N_2(l) \geqq 2$. q.e.d.

Combining (19-54) with (19-48), we obtain six inequalities

$$M_a(\bar{G}) + M_b(\bar{G}) \leqq M_{ac}(\bar{G}) + M_{ad}(\bar{G}), \quad \text{etc.} \tag{19-57}$$

which are always valid under Assumption 19-4. Thus we no longer need *ad hoc* assumptions such as (19-38). Hence instead of (19-36) we obtain that $\bar{\varphi}_{G,ad} = 0$ unless

$$\begin{aligned} \varkappa \geqq\ & z_a[M_a(\bar{G})]^2 + z_b[M_b(\bar{G})]^2 + z_c[M_{ac}(\bar{G}) - M_a(\bar{G})]^2 \\ & + z_d[M_b(\bar{G}) - M_{ac}(\bar{G})]^2 + z_{ab}[M_a(\bar{G}) - M_b(\bar{G})]^2 \\ & + z_{ac}[M_{ac}(\bar{G})]^2, \quad \text{etc.} \end{aligned} \tag{19-58}$$

Hence Theorem 19-5 yields that $\varphi_{ad} = 0$ unless

$$\begin{aligned} \varkappa \geqq \max\,[&\hat{z}_a + \hat{z}_b + \hat{z}_{ac},\ \hat{z}_c + \hat{z}_d + \hat{z}_{ac}, \\ &\hat{z}_a + \hat{z}_d + \hat{z}_{ac},\ \hat{z}_b + \hat{z}_c + \hat{z}_{ac}, \\ &\hat{z}_a + \hat{z}_c + \hat{z}_{ab},\ \hat{z}_b + \hat{z}_d + \hat{z}_{ab}, \\ &\hat{z}_a + \hat{z}_d + \hat{z}_{ab},\ \hat{z}_b + \hat{z}_c + \hat{z}_{ab}] \end{aligned} \tag{19-59}$$

with $\hat{z}_h \equiv z_h M_h^2$ $(h = a, b, c, d, ab, ac)$. For example, from (19-59) we have

$$\varkappa \geqq \tfrac{1}{2} \sum_{\nu=a}^{d} z_\nu M_\nu^2 + \max\,(z_{ab}M_{ab}^2, z_{ac}M_{ac}^2) \tag{19-60}$$

§ 20 MAJORIZATION METHOD

20-1 Models

As explained in Subsection 6-1, hadrons are only particles which strongly interact each other. If other interactions are neglected, several hadrons are regarded as stable. Those stable hadrons have particular masses and specific quantum mumbers. We construct some models by abstracting the following qualitative features from their properties.

First of all, there exist particles having the smallest mass, which are called *pions*. Their mass is taken to be unity. We sometimes require pions to satisfy the parity conservation law; its definition is given in Subsection 20-3. We call G an *equal-mass Feynman graph* if all particles involved in G are pions and if the parity conservation law is not taken into account.

In general, we also have various mesons and baryons (pions are a kind of mesons). Each baryon line has an *intrinsic* orientation, that is, for any line associated with a baryon we cannot arbitrarily assign outgoing and ingoing to its two end points. If a Feynman graph G involves baryon lines, we require G to satisfy the *baryon-number conservation law*, that is, for any vertex of G, an outgoing baryon line and an ingoing one (without distinguishing external lines from internal ones) can be incident with it only in a pair (or pairs). This rule immediately implies that the number r of the outgoing baryon external lines of G is equal to that of the ingoing ones, whence $2r \leqq n$. We shall below define a set, Ψ_n^r, of such Feynman graphs. The baryon-number conservation law also implies the existence of r disjoint baryon paths, where a *baryon path* is a path consisting of baryon lines alone between two external vertices, one of which an outgoing baryon external line is incident with and the other of which an ingoing one is incident with.

Experimentally, there is another quantum number, called strangeness, which conserves in the strong interactions, but here we do not take the strangeness conservation law into account.

The nucleons are the baryons of the smallest mass, which is denoted by m hereafter in this chapter; of course $m \geqq 1$.

As emphasized in Subsection 17-1, the existence of the baryon-number conservation law has an important effect on the supports of weight distributions. The following example explains how Ψ_3^1 is different from Φ_3 when they satisfy the same spectral conditions.

Example We consider the condition for the absence of an anomalous threshold of the "nucleon vertex function" in the triangle graph shown in Fig. 13-4; the particles of the external lines incident with a and b are nucleons. For simplicity, we put $m_1 = m_2$ and $s_a = s_b \leqq (m_1 + m_3)^2$. From (13-27), the condition for no anomalous threshold in s_c reads

$$s_a \leqq m_1^2 + m_3^2 \tag{20-1}$$

If the baryon-number conservation law is taken into account, all the minima of $m_1 + m_3$, $m_2 + m_3$, and $m_1 + m_2$ are realized in the case in which the line 3 is of a nucleon and the lines 1 and 2 are of pions. Then the condition (20-1) reduces to $s_a \leqq m^2 + 1$; the nucleon's mass shell $s_a = m^2$ is allowed for any value of m. Since the above identification of particles implies $M_a = M_b = m + 1$ and $M_c = 2$, if we impose *only* the spectral conditions with those values of M_a, M_b, M_c on the internal masses, we can choose $m_1 = m_2 = m_3 = (m + 1)/2$. Then (20-1) becomes $s_a \leqq (m + 1)^2/2$; the mass shell is allowed only for $m \leqq \sqrt{2} + 1$. Thus if m is much larger than unity as observed experimentally ($m \doteqdot 6.7$), the baryon-number conservation law is very important.

Now, we have to define the set Ψ_n^r precisely. If we strictly adhere to the Lagrangian field theory, we shall be bothered by inessential complications. For example, a particular interaction Lagrangian L_{int} admits only particular types of vertices, but there is no inevitability of choosing L_{int}. To be general as well as possible, we will not specify the types of vertices more than the baryon-number conservation law.

DEFINITION **20-1** The set Ψ_n^r consists of all *strongly connected* Feynman graphs G satisfying the following conditions:

(a) $m_l \geqq 1$ for any $l \in |G|$.

(b) The baryon-number conservation law is satisfied.

(c) G has $2r$ baryon external lines and $n - 2r$ meson ones.

It is important to note that if $G \in \Psi_n^r$ then $G/I \in \Psi_n^r$ for any $I \subset |G|$. According to Definition 20-1, Ψ_n^r does not contain the Feynman graphs which have cut-lines. We have excluded cut-lines because since they induce only poles, it is easy and convenient to take them into account later on. [We do not explicitly describe them any more, however.] For any G' satisfying (a), (b), (c) of Definition 20-1, Ψ_n^r does contain G'/I' instead of G', where I' is the set of all cut-lines of G'.

The problem of discussing

$$\text{Supp}\,[\Psi_n^r] \equiv \bigcup_{G \in \Psi_n{}^r} \text{supp}\,\bar{\varphi}_G \tag{20-2}$$

can be reduced to finding $p = \bar{p}$ at which (19-2) holds, as in Subsection 19-1. It is no longer good, however, to consider the case in which all of $\bar{p}_b$ ($b \in g$) are parallel, because then no characteristics of Ψ_n^r can be taken into account. If we still wish to apply Theorem 19-1 to our problem, we have to solve a non-linear problem, which is beyond the scope of the theory of linear programming. We quote here the following theorem, which can be proven by means of the path theorem (Theorem 5-4). We call the method based on the path theorem the *path method.*

THEOREM **20-2** [10] If G is a strongly connected equal-mass Feynman graph with $n = 4$, we can find q_l ($l \in |G|$) satisfying the conditions of Theorem 19-1 for the euclidean external momenta defined by

$$\begin{aligned} p_a &= -k_1 - k_2, \quad & p_b &= -k_1 + k_2 \\ p_c &= k_1 - k_2, \quad & p_d &= k_1 + k_2 \end{aligned} \tag{20-3}$$

where $k_1^2 = k_2^2 = 1$ and $k_1 k_2 = 0$.

This theorem is particularly useful for the case in which all four external particles are of the same kind (e.g., Ψ_4^2). In the other cases, it is better to employ majorization methods, where a *majorization method* is a method of reducing the inequalities (19-2) for given complicated graphs to those for very simple ones. Two majorization methods are known: one is called *euclidean* [28, 25] and the other is *noneuclidean* [29, 30] according as $\{\bar{p}_b \mid b \in g\}$ is restricted to a euclidean set or not. Generally speaking, the latter method, which we describe in the next subsection in detail, is very powerful for the transition amplitudes but not for the quantities off the mass shell. We omit the euclidean majorization method and quote its results only when it yields better results than those which can be obtained by the noneuclidean majorization method.

20-2 Majorization Rules

In this subsection, we present some basic majorization rules of the noneuclidean majorization method. Since we consider the V functions corresponding to various Feynman graphs, we write the V of G as $V_G(p, \alpha)$ and the Δ of G as Δ_G.

DEFINITION **20-3** If

$$V_G(p, \alpha) \geqq 0 \quad \text{for all} \quad \alpha \in \Delta_G \tag{20-4}$$

follows from the assumption that

$$V_{G_j}(p, \alpha) \geqq 0 \quad \text{for all} \quad \alpha \in \Delta_{G_j} \tag{20-5}$$

for $j = 1, \ldots, k$, then G is said to be *majorized* by the Feynman graphs $G_1, \ldots, G_k$, and the majorization is denoted by

$$G \to \{G_1, \ldots, G_k\} \tag{20-6}$$

For $k = 1$, we omit the curly bracket in (20-6).

The following theorem is a direct consequence of Definition 20-3.

THEOREM **20-4** If there exist some $j\,(1 \leqq j \leqq k)$, some $\alpha \in \Delta_{G_j}$, and some $c > 0$ such that

$$cV_G(p, \bar{\alpha}) \geqq V_{G_j}(p, \alpha) \tag{20-7}$$

where $V_G(p, \alpha)$ with p fixed assumes the minimum value at $\alpha = \bar{\alpha}$ in $\Delta_G^\lambda \equiv \left\{\alpha \mid \alpha_l \geqq 0 \text{ for all } \alpha_l, \sum_{l \in |G|} \alpha_l = \lambda > 0\right\}$, then G is majorized by $G_1, \ldots, G_k$.

THEOREM **20-5** Let $(Q_G)_l$ be the increment (with p and $\alpha_{l'}$ for all $l' \neq l$ fixed) of

$$Q_G \equiv \sum_{k \in |G|} \alpha_k m_k^2 - V_G \tag{20-8}$$

when α_l is increased by δ_l (not infinitesimal). Then

$$(Q_G)_l = \frac{\varrho_l \delta_l}{\varrho_l + \delta_l} Y_l^2 \tag{20-9}$$

where

$$\varrho_l \equiv \frac{U}{\partial U / \partial \alpha_l} \tag{20-10}$$

and Y_l is given in Subsection 9-2.

Proof According to the circuit representation (7-26) of V, we have

$$Q_G = \sum_k \alpha_k q_k^2 - \sum_C (U_C/U) \left(\sum_k [C:k]\, \alpha_k q_k\right)^2 \tag{20-11}$$

We can choose $q_k\,(k \in |G|)$ in such a way that all the circuit equations (12-6) are satisfied identically, that is, we set $q_k = Y_k\,(k \in |G|)$ [see (12-11)]. Using the fact that U and U_c are linear in α_l, from (20-11) we find

$$(Q_G)_l = \delta_l Y_l^2 - \sum_C \frac{U_C + \delta_l\, \partial U_C / \partial \alpha_l}{U + \delta_l\, \partial U / \partial \alpha_l} [C:l]^2\, \delta_l^2 Y_l^2 \tag{20-12}$$

By means of (7-33) and $\partial U_C/\partial\alpha_l = 0$ for $l \in C$, (20-12) is reduced to

$$(Q_G)_l = \frac{U\,\delta_l Y_l^2}{U + \delta_l\,\partial U/\partial\alpha_l} \tag{20-13}$$

q.e.d.

THEOREM **20-6** If l is not a cut-line,

$$Q_{G-l} - Q_G = \varrho_l Y_l^2 \tag{20-14}$$

If l is a cut-line, of course the left-hand side of (20-14) equals $-\alpha_l q_l^2$.

Proof We note from (7-45) that

$$\lim_{\alpha_l\to\infty} Q_G = Q_{G-l} \tag{20-15}$$

because the linearity in α_l implies

$$\lim_{\alpha_l\to\infty} \frac{W^{(b|c)}}{U} = \frac{\partial W^{(b|c)}/\partial\alpha_l}{\partial U/\partial\alpha_l} \tag{20-16}$$

and if l is not a cut-line then $\partial U/\partial\alpha_l$ and $\partial W^{(b|c)}/\partial\alpha_l$ are the U and the $W^{(b|c)}$ in $G - l$, respectively, according to Theorem 3-19. Therefore, (20-9) with $\delta_l \to \infty$ yields (20-14). q.e.d.

We rewrite (20-14) as

$$V_G(p, \alpha) = V_{G-l}(p, \alpha) + \alpha_l m_l^2 + \varrho_l Y_l^2 \tag{20-17}$$

If $V_G(p, \alpha)$ assumes the minimum at $\alpha = \bar{\alpha}$ in Δ_G^λ, we should have either $\bar{\alpha}_l = 0$ or

$$m_l^2 - Y_l^2\big|_{\alpha=\bar{\alpha}} = \frac{\partial V_G}{\partial\alpha_l}\bigg|_{\alpha=\bar{\alpha}} = \lambda^{-1} V_G(p, \bar{\alpha}) \tag{20-18}$$

according to (9-31) and the method of Lagrange multiplier. We therefore have *either*

$$V_G(p, \bar{\alpha}) = V_{G/l}(p, \alpha) \qquad (\bar{\alpha}_l = 0) \tag{20-19}$$

because of Theorem 9-1, *or*, from (20-17) and (20-18),

$$c_l V_G(p, \bar{\alpha}) = V_{G-l}(p, \bar{\alpha}) + (\bar{\alpha}_l + \bar{\varrho}_l)\, m_l^2 \tag{20-20}$$

where $\bar{\varrho}_l \equiv \varrho_l|_{\alpha=\bar{\alpha}}$ and $c_l \equiv 1 + \lambda^{-1}\,\bar{\varrho}_l > 0$ (with $\lambda = 1 + \bar{\alpha}_l$).

After the above preparation, we consider majorization rules. When G is majorized by $G_1, \ldots, G_k$, we usually require that

(a) if $G \in \Psi_n^r$ then $G_j \in \Psi_n^r$ for $j = 1, \ldots, k$, and

(b) $N(G) > N(G_j)$ for $j = 1, \ldots, k$.

The majorization is called *admissible* if (a) is satisfied and *normal* if (b) is satisfied. If each of $G_1, \ldots, G_k$ is further majorized by some among $G'_1, \ldots, G'_m$, then evidently G is majorized by $G'_1, \ldots, G'_m$, that is, majorizations can be *combined*. This combined majorization is admissible and normal if all the constituent majorizations are admissible and normal, respectively. Thus by combining many admissible, normal majorizations, we can reduce a given Feynman graph G into a set of much simpler graphs.

Majorization rules are general ways of majorizing the Feynman graphs which have certain characteristics. Majorization rules should usually give us admissible, normal majorizations, but we sometimes use nonadmissible or abnormal ones. A nonadmissible majorization has to be followed by other nonadmissible ones in such a way that the combined majorization is admissible. Abnormal majorizations are used only in the initial and final stages of majorizations to simplify the descriptions. The following auxiliary majorization rules give us abnormal majorizations (except for $G \to \hat{G}$).

THEOREM **20-7** Let G be an arbitrary Feynman graph.

(A1) $G \to \hat{G}$, where $\hat{G}$ is G with some internal masses reduced in value.

(A2) $G \to G'$ and $G' \to G$, where G' is obtained from G by replacing some parallel lines in G by a single line whose mass equals the sum of theirs.

(A3) $G/l \to G$ for any $l \in |G|$.

Rules (A1) and (A2) are not necessarily admissible, but it is easy to find in what cases they are admissible.

Proof (A1). Evident. (A2). We use the inverse-Feynman-parametric representation. If l and l' are parallel lines, Q depends on β_l and $\beta_{l'}$ only through $\beta_l + \beta_{l'} \equiv \beta_{ll'}$. Furthermore,

$$\inf_{\beta_l + \beta_{l'} = \beta_{ll'}} \left(\frac{m_l^2}{\beta_l} + \frac{m_{l'}^2}{\beta_{l'}} \right) = \frac{(m_l + m_{l'})^2}{\beta_{ll'}} \tag{20-21}$$

Thus $\inf V_G = \inf V_{G'}$. (A3). This is evident from $V_{G/l} = V_G|_{\alpha_l = 0}$. q.e.d.

Now, we state main majorization rules, which give us normal majorizations.

THEOREM **20-8** [29, 30] Let G be an arbitrary Feynman graph belonging to Ψ_n^r.

(M1) $G \to G/l$ if there exists an internal vertex a such that $S[a] = \{l, l'\}$ and $m_l \geqq m_{l'}$.

(M2) $G \to \{G_1, G_2\}$ if G is separable at a cut-vertex a into $G_1 = G/|G_2|$ and $G_2 = G/|G_1|$.

(M3) $G \to \{G/l, G - l\}$ for any $l \in |G|$, but this majorization is admissible only when $G - l$ is strongly connected and l is not a baryon line.

(M4) $G \to \{G/l, (G - l)/l'\}$ if there exists an external vertex a such that $S[a] = \{l, l'\}$ and if $s_a \leqq m_l^2 + m_{l'}^2$, (see Fig. 20-1). This majorization is admissible only when $(G - l)/l'$ is strongly connected and l is not a baryon line.

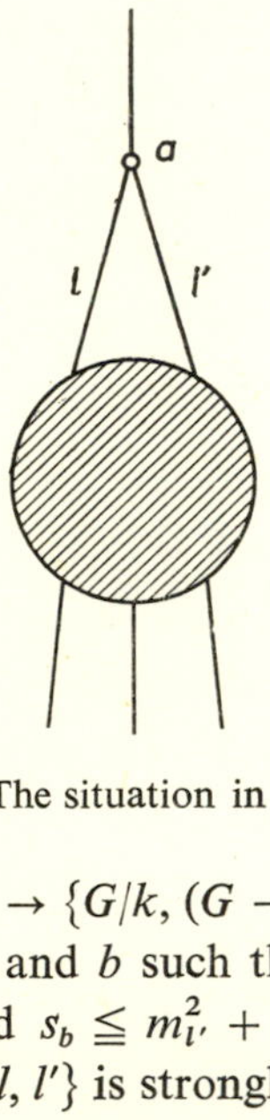

Fig. 20-1 The situation in Rule (M4).

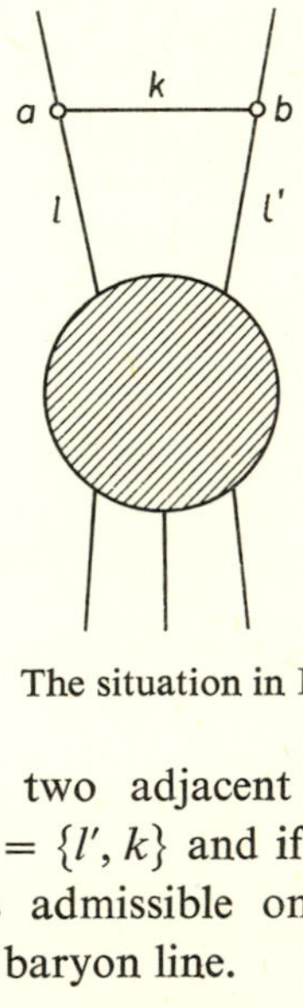

Fig. 20-2 The situation in Rule (M5).

(M5) $G \to \{G/k, (G - k)/\{l, l'\}\}$ if there exist two adjacent external vertices a and b such that $S[a] = \{l, k\}$ and $S[b] = \{l', k\}$ and if $s_a \leqq m_l^2 + m_k^2$ and $s_b \leqq m_{l'}^2 + m_k^2$. This majorization is admissible only when $(G - k)/\{l, l'\}$ is strongly connected and k is not a baryon line.

Proof: (M1) We observe that Q_G depends on α_l and $\alpha_{l'}$ only through $\alpha_l + \alpha_{l'}$. (M2). This follows from $V_G = V_{G_1} + V_{G_2}$. (M3). This is a direct consequence of (20-19) and (20-20) according to Theorem 20-4.

(M4). Since (20-19) gives us G/l, we have only to consider (20-20). Since l' is a cut-line in $G - l$, we have

$$V_{G-l} = V_{(G-l)/l'} + \alpha_{l'}(m_{l'}^2 - s_a) \qquad (20\text{-}22)$$

Therefore, from (20-20) together with $s_a \leqq m_l^2 + m_{l'}^2$, we find

$$c_l V_G(p, \bar{\alpha}) \geqq V_{(G-l)/l'}(p, \bar{\alpha}) + (\bar{\alpha}_l + \bar{\varrho}_l - \bar{\alpha}_{l'})\, m_l^2 \tag{20-23}$$

Since, in G, U depends on α_l and $\alpha_{l'}$ only through $\alpha_l + \alpha_{l'}$, we see

$$U \geqq (\alpha_l + \alpha_{l'})\, \partial U / \partial \alpha_l \tag{20-24}$$

Hence (20-10) implies

$$\bar{\varrho}_l \geqq \bar{\alpha}_l + \bar{\alpha}_{l'} \tag{20-25}$$

Thus the last term of (20-23) is positive.

(M5). We have only to consider the case $\bar{\alpha}_k \neq 0$ as before. Since

$$V_{G-k} = V_{(G-k)/\{l,l'\}} + \alpha_l(m_l^2 - s_a) + \alpha_{l'}(m_{l'}^2 - s_b) \tag{20-26}$$

(20-10) implies

$$\begin{aligned} c_k V_G(p, \bar{\alpha}) &- V_{(G-k)/\{l,l'\}} \\ &= (\bar{\alpha}_k + \bar{\varrho}_k)\, m_k^2 + \bar{\alpha}_l(m_l^2 - s_a) + \bar{\alpha}_{l'}(m_{l'}^2 - s_b) \\ &\geqq \bar{\alpha}_l(m_l^2 + m_k^2 - s_a) + \bar{\alpha}_{l'}(m_{l'}^2 + m_k^2 - s_b) + 2\bar{\alpha}_k m_k^2 \\ &\geqq 0 \end{aligned} \tag{20-27}$$

with the aid of

$$\bar{\varrho}_k \geqq \bar{\alpha}_l + \bar{\alpha}_{l'} + \bar{\alpha}_k \tag{20-28}$$

q.e.d.

20-3 Majorization Procedure

According to Theorem 20-8, Majorization Rules (M4) and (M5) do not allow external masses to become larger than certain values. Hence we should consider the transition amplitudes rather than the extended ones. For simplicity of description, we assume that any external line is of either a pion or a nucleon, but we may replace it by a meson or a baryon if its mass squared is not greater than 2 (for a meson) or $m^2 + 1$ (for a baryon).

We first consider Ψ_n^0, for which all external lines are of pions. We assume $s_a \leqq 2$ for any $a \in g$. By means of Rule (A1), we may put $m_l = 1$ for all $l \in |G|$, that is, we have only to consider the equal-mass Feynman graphs.

Let Ω_n' be the set of all equal-mass Feynman graphs which cannot be majorized by others of Ψ_n^0 by means of Rules (M1)–(M5).

Any graph $G \in \Omega_n'$ is nonseparable because of Rule (M2) and has no internal vertex of degree 2 because of Rule (M1). We call a vertex a such that $N(S[a]) \geqq 3$ a *node*. [Note that we do not count the number of external lines incident with a.] We also call a path between two distinct nodes which passes through no other nodes an *n-path*. Since G has no internal vertex of degree 2, on each n-path there can lie external vertices only.

If G has an n-path P on which less than three external vertices lie, we can majorize G by means of one of Rules (M3), (M4), and (M5) if this majorization is admissible. It is not admissible only when there is another n-path P' such that $S \equiv \{l, l'\}$ is a cut-set for any $l \in P$ and any $l' \in P'$. Let H and H' be two connected subgraphs of $G - S$. Since the end points of P and P' are nodes, H contains at least two nodes because if P and P' had a common node then it would be a cut-vertex, and moreover H includes a circuit which passes through at least two nodes, because they are not cut-vertices. This circuit is composed of at least two n-paths, on any of which there lie at least three external vertices because otherwise G is majorized in the admissible way by means of one of Rules (M3), (M4), and (M5). Therefore H contains at least six external vertices. The same is true also for H'. Hence $n \geqq 12$.

Thus, for $n \leqq 11$, on every n-path in G there lie at least three external vertices. Hence for $n \leqq 8$ G can have no node, that is, it is a single-loop graph. For $9 \leqq n \leqq 11$, some two-node graphs are possible. Let Ω_n be the subset of Ω_n' consisting of the Feynman graphs having n distinct external vertices which are also distinct from nodes. Then any graph of $\Omega_n' - \Omega_n$ is majorized by a graph of Ω_n by repeated use of Rule (A3). Thus we obtain the following theorem.

THEOREM **20-9** [29]† If $s_a \leqq 2$ for any $a \in g$, then any graph of Ψ_n^0 is majorized by the graphs of Ω_n, where Ω_n for $n \leqq 8$ consists of single-loop Feynman graphs having n distinct external vertices alone, and Ω_9 and Ω_{10},

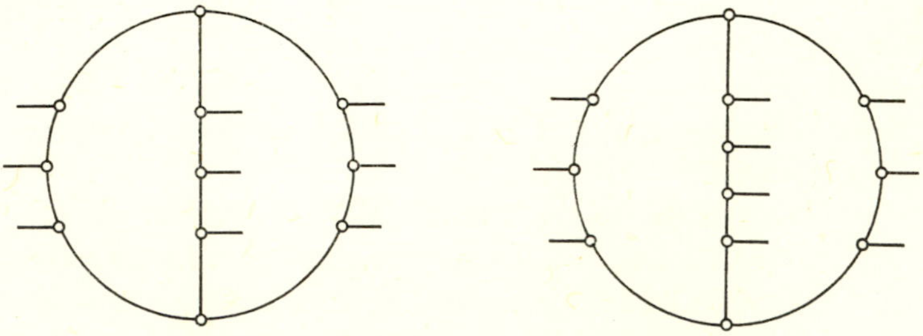

Fig. 20-3 (a) & (b) The non-single-loop graphs of Ω_9 and Ω_{10}.

† The above explicit proof was not presented in the original paper.

in addition to those graphs, contain the graphs shown in Fig. 20-3 (a) and (b), respectively.

It can be shown that for $n \geqq 11$ the domain in which $V_G \geqq 0$ for all $G \in \Psi_n^0$ is empty even for $s_a = 1$ for all $a \in g$ [29].

Next, we consider Ψ_n^r with $n/2 \geqq r \geqq 1$. By means of Rule (A1), all baryon and meson internal lines can be replaced by nucleon and pion internal lines, respectively.

In the present case, the majorizations given by Rules (M3), (M4), and (M5) are not admissible when a nucleon line is deleted, because then the baryon-number conservation law is violated. It turns out to be convenient, however, to use such nonadmissible majorizations sometimes. Furthermore, the majorization rules presented in Theorem 20-8 are not sufficient for the present purpose, that is, one has to add some other majorization rules. Then, according to Boyling [30], one can explicitly carry out the majorization of Ψ_n^r for $n \leqq 6$ under the restrictions that $s_a \leqq 2$ for all pion external lines, $s_b \leqq m^2 + 1$ for all nucleon ones, and $s_h \leqq M_h^2$ for all other invariant squares. Any graph of Ψ_n^r is majorized by the graphs of Ω_n^r, where Ω_4^r and Ω_5^r consist of single-loop graphs alone but Ω_6^r contains several topologically different graphs. For example, Ω_4^1 and Ω_4^2 consist of the graphs shown in Fig. 20-4 and in Fig. 20-5, respectively.

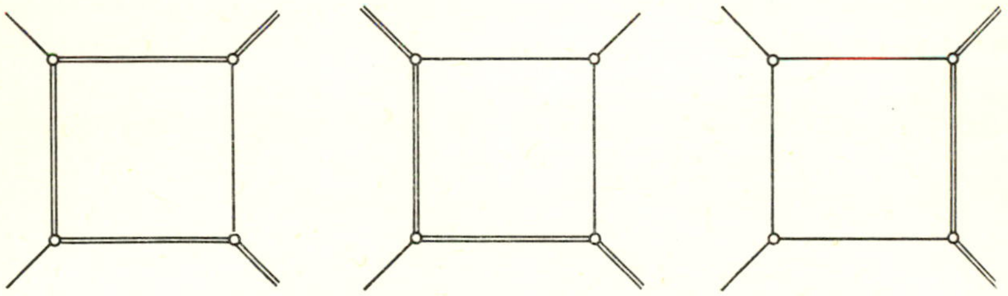

Fig. 20-4 The graphs of Ω_4^1. Double lines stand for nucleon lines.

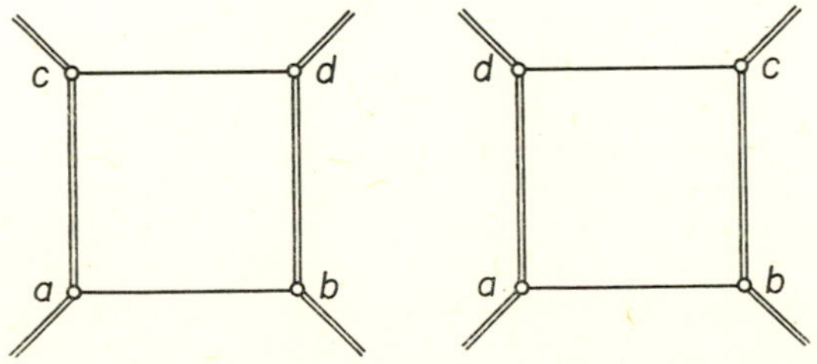

Fig. 20-5 The graphs of Ω_4^2. Double lines stand for nucleon lines.

For $r \geqq 2$, some care has to be taken. For example, consider Ψ_4^2, the nucleon-nucleon scattering. According to the above majorization, every graph G of Ψ_4^2 is majorized by a pair of the graphs shown in Fig. 20-5. Actually, it turns out, however, that G is majorized by *either* of those two graphs alone. This latter majorization is called the *classified majorization*, and correspondingly the former is called the *unclassified majorization*. Although both majorizations yield the same Symanzik region D_0 [see (18-29)], the classified majorization leads us to better support properties for PTIR. For example, for (17-22) we find that $\varrho_{tu}(z, \gamma)$ vanishes unless $\gamma \geqq 4$ if the unclassified majorization is employed, and that it vanishes unless

$$\gamma \geqq \min\,[4m^2 z + 4(1 - z), \quad 4z + 4m^2(1 - z)] \tag{20-29}$$

if the classified one is employed.

Finally, we briefly mention the *parity conservation law*. Physically, this law implies that the strong interactions are invariant under the space reflection. For the present purpose, we can state it as follows: If pion lines (including external lines) only are incident with a vertex of G or of any reduced graph of G, then their number (i.e., the F-degree of that vertex) has to be even; if there are mesons other than pions, the above law should be applied after replacing each of them by two or three pions appropriately in the sense of Rules (A1) and (A2), where we assume that the masses of the particles other than pions are not smaller than 2. We note that the circuits formed by baryon lines alone (called *baryon loops*) can consistently be replaced by the one formed by pion lines in the sense of Rules (A1) and (A2). In this way, we have graphs such that any vertex which does not lie on any baryon path has an even F-degree. We denote by $\bar{\Psi}_n^r$ the subset of Ψ_n^r consisting of all such graphs.

It is very difficult to discuss $\bar{\Psi}_n^r$ in general because Rule (M3) usually violates the parity conservation law. The path method is more convenient to discuss $\bar{\Psi}_n^0$ (n even).

§ 21 W FUNCTION INEQUALITIES

In this section, we consider the case $n = 4$ in more detail. As a result of the majorization procedure presented in Subsection 20-3, in order to show $V \geqq 0$ for all graphs of Ψ_4^r, we have only to investigate the square graph, namely, the $N = 4$ single-loop graph, provided that

$$s_\nu \leqq m_\nu^2 + 1 \qquad (\nu = a, b, c, d) \tag{21-1}$$

where a, b, c, d are the four external vertices and $m_\nu = 1$ or m according as the corresponding external line is of a pion or of a nucleon. Since, as stated in Subsection 13-2, no anomalous thresholds are present in the square graphs under (21-1), we have $V \geqq 0$ for any graph G of Ψ_4^r if s, t, u are not greater than the normal thresholds M_{ab}^2, M_{ac}^2, M_{ad}^2, respectively. The Symanzik region

$$D_0 = \{s, t \mid s < M_{ab}^2, t < M_{ac}^2, u < M_{ad}^2\} \tag{21-2}$$

is nonempty for $s_\nu = m_\nu^2 + 1$ $(\nu = a, b, c, d)$ if

$$M_{ab}^2 + M_{ac}^2 + M_{ad}^2 < \sum_{\nu=a}^{d} m_\nu^2 + 4 \tag{21-3}$$

This inequality holds for any of Ψ_4^r $(r = 0, 1, 2)$. Hence for $s_\nu = m_\nu^2 + 1$ $(\nu = a, b, c, d)$, $s = M_{ab}^2$, and $t = M_{ac}^2$, we have

$$V = \sum_l \alpha_l m_l^2 - \sum_{\nu=a}^{d} (\zeta_\nu + \zeta_{ad})(m_\nu^2 + 1)$$
$$- (\zeta_{ab} - \zeta_{ad}) M_{ab}^2 - (\zeta_{ac} - \zeta_{ad}) M_{ac}^2 \geqq 0 \tag{21-4}$$

namely,

$$\sum_l \alpha_l m_l^2 - \sum_{\nu=a}^{d} (\zeta_\nu + \zeta_{ad}) m_\nu^2$$
$$\geqq (\zeta_{ab} - \zeta_{ad}) M_{ab}^2 + (\zeta_{ac} - \zeta_{ad}) M_{ac}^2 + \sum_{\nu=a}^{d} (\zeta_\nu + \zeta_{ad}) \tag{21-5}$$

In Δ_{ad} [see (17-18)], we obtain from (21-5) a support property of $\varrho_{st}(z, \gamma)$ of (17-22),

$$\gamma \geqq zM_{ab}^2 + (1 - z) M_{ac}^2 \tag{21-6}$$

with the aid of (17-23). In (21-6), however, the last term of the right-hand side of (21-5) is wasted. In order to utilize it fully, we need some inequalities†

† Those inequalities were proposed by Nakanishi [12] and proven by Boyling [31].

among ζ functions. In this and later sections, we write W_h instead of $W^{(h|g-h)}$ for simplicity. Since $\zeta_h = W_h/U$, we should find inequalities among W functions or $\tilde{W}$ functions (see Subsection 3-2).

THEOREM **21-1** [31] The $\tilde{W}$ functions satisfy a quadratic identity

$$(\tilde{W}_a + \tilde{W}_{ad})(\tilde{W}_d + \tilde{W}_{ad}) - (\tilde{W}_{ab} - \tilde{W}_{ad})(\tilde{W}_{ac} - \tilde{W}_{ad}) = \tilde{U}\tilde{W}_{ad}^{(ad)} \tag{21-7}$$

where $\tilde{W}_{ad}^{(ad)}$ denotes the $\tilde{W}_{ad}$ function in $G^{(ad)}$, the graph obtained from G by identifying a and d.

Proof From Theorems 3-16 and 3-18 together with (3-45), we have

$$\begin{aligned}
[\mathscr{W}^a]^{d,d} &= \tilde{W}^{(a|d)} = \tilde{W}_a + \tilde{W}_d + \tilde{W}_{ab} + \tilde{W}_{ac} \\
[\mathscr{W}^a]^{b,c} &= \tilde{W}^{(a|bc)} = \tilde{W}_a + \tilde{W}_{ad} \\
[\mathscr{W}^a]^{b,d} &= \tilde{W}^{(a|bd)} = \tilde{W}_a + \tilde{W}_{ac} \\
[\mathscr{W}^a]^{c,d} &= \tilde{W}^{(a|cd)} = \tilde{W}_a + \tilde{W}_{ab}
\end{aligned} \tag{21-8}$$

It is straightforward to rewrite (21-8) as

$$\begin{aligned}
\tilde{W}_a + \tilde{W}_{ad} &= [\mathscr{W}^a]^{b,c} \\
\tilde{W}_d + \tilde{W}_{ad} &= [\mathscr{W}^a]^{d,d} + [\mathscr{W}^a]^{b,c} - [\mathscr{W}^a]^{b,d} - [\mathscr{W}^a]^{c,d} \\
\tilde{W}_{ab} - \tilde{W}_{ad} &= [\mathscr{W}^a]^{c,d} - [\mathscr{W}^a]^{b,c} \\
\tilde{W}_{ac} - \tilde{W}_{ad} &= [\mathscr{W}^a]^{b,d} - [\mathscr{W}^a]^{b,c}
\end{aligned} \tag{21-9}$$

Hence

$$\begin{aligned}
K &\equiv (\tilde{W}_a + \tilde{W}_{ad})(\tilde{W}_d + \tilde{W}_{ad}) - (\tilde{W}_{ab} - \tilde{W}_{ad})(\tilde{W}_{ac} - \tilde{W}_{ad}) \\
&= [\mathscr{W}^a]^{b,c}[\mathscr{W}^a]^{d,d} - [\mathscr{W}^a]^{b,d}[\mathscr{W}^a]^{c,d}
\end{aligned} \tag{21-10}$$

Because of Jacobi's theorem for determinants, (21-10) becomes†

$$\begin{aligned}
K &= (\det \mathscr{W}^a)[\mathscr{W}^a]^{bd,cd} \\
&= \tilde{U}[\mathscr{W}^{ad}]^{b,c} = \tilde{U}\tilde{W}_{ad}^{(ad)}
\end{aligned} \tag{21-11}$$

because $\tilde{W}_a^{(ad)} \equiv 0$. q.e.d.

Since the right-hand side of (21-7) is nonnegative, we obtain an inequality

$$(W_a + W_{ad})(W_d + W_{ad}) \geqq (W_{ab} - W_{ad})(W_{ac} - W_{ad}) \qquad \text{for all} \quad \alpha_l \geqq 0 \tag{21-12}$$

† $\mathscr{W}^{ad}$ denotes the matrix $\mathscr{W}^a$ in $G^{(ad)}$.

in the Feynman-parametric form. Hence

$$(W_a + W_d + 2W_{ad})^2 \geqq 4(W_{ab} - W_{ad})(W_{ac} - W_{ad}) \tag{21-13}$$

In Δ_{ad}, therefore, we find

$$\zeta_a + \zeta_d + 2\zeta_{ad} \geqq 2[(\zeta_{ab} - \zeta_{ad})(\zeta_{ac} - \zeta_{ad})]^{\frac{1}{2}} \tag{21-14}$$

and likewise

$$\zeta_b + \zeta_c + 2\zeta_{ad} \geqq 2[(\zeta_{ab} - \zeta_{ad})(\zeta_{ac} - \zeta_{ad})]^{\frac{1}{2}} \tag{21-15}$$

Thus (21-5) together with (17-23) yields that $\varrho_{st}(z, \gamma)$ vanishes unless

$$\gamma \geqq zM_{ab}^2 + (1 - z)M_{ac}^2 + 4[z(1 - z)]^{\frac{1}{2}} \tag{21-16}$$

instead of (21-6).

REMARK **21-2** If we identify b and c so that $\tilde{W}_{ab} = \tilde{W}_{ac} = 0$, Theorem 21-1 reduces to

$$\tilde{W}_a\tilde{W}_d + \tilde{W}_d\tilde{W}_{ad} + \tilde{W}_{ad}\tilde{W}_a = \tilde{U}\tilde{W}_{ad}^{(ad)} \tag{21-17}$$

Since we now have a Feynman graph with $n = 3$, we relabel the external vertices as $a' = a$, $b' = d$, and $c' = b = c$. Then (21-17) is rewritten as

$$\tilde{W}_{a'}\tilde{W}_{b'} + \tilde{W}_{b'}\tilde{W}_{c'} + \tilde{W}_{c'}\tilde{W}_{a'} = \tilde{U}\tilde{W}_0 \tag{21-18}$$

where $\tilde{W}_0$ denotes the 3-tree product sum with respect to a', b', c'. [A 3-tree with respect to a', b', c' is a tree in the graph obtained by identifying a', b', c'.] It is interesting to note that (21-18) presents an *exact* relation between the numbers of trees, 2-trees, and 3-trees in an arbitrary graph.

If we confine ourselves to the application to the support properties of PTIR, Theorem 21-1 is sufficient. The remainder of this section is devoted to the problems of mathematical interest only.

First, the inequality (21-12) can be made more precise.

THEOREM **21-3** [12, 31]

$$\begin{aligned} W_aW_d &\geqq W_{ab}W_{ac}, & W_bW_c &\geqq W_{ab}W_{ac} \\ W_aW_b &\geqq W_{ac}W_{ad}, & W_cW_d &\geqq W_{ac}W_{ad} \\ W_aW_c &\geqq W_{ad}W_{ab}, & W_bW_d &\geqq W_{ad}W_{ab} \\ &\text{for all} \quad \alpha_l \geqq 0 \end{aligned} \tag{21-19}$$

Proof Of course we have only to prove the first inequality of (21-19). In the inverse-Feynman-parametric form, we find

$$\begin{aligned} \tilde{W}_a\tilde{W}_d - \tilde{W}_{ab}\tilde{W}_{ac} &= (\tilde{W}_a + \tilde{W}_{ad})(\tilde{W}_d + \tilde{W}_{ad}) - (\tilde{W}_{ab} - \tilde{W}_{ad})(\tilde{W}_{ac} - \tilde{W}_{ad}) \\ &\quad - \tilde{W}_{ad}(\tilde{W}_a + \tilde{W}_d + \tilde{W}_{ab} + \tilde{W}_{ac}) \\ &= \tilde{U}\tilde{W}_{ad}^{(ad)} - \tilde{W}_{ad}\tilde{U}^{(ad)} \end{aligned} \tag{21-20}$$

from (21-7) and the first formula of (21-8) together with

$$\tilde{U}^{(ad)} \equiv \tilde{W}^{(a|d)} \tag{21-21}$$

where $\tilde{U}^{(ad)}$ denotes the $\tilde{U}$ function in $G^{(ad)}$.

For $S \in \boldsymbol{S}(ad \mid bc)$, let H and H' be the two connected subgraphs of $G - S$, where H has the two external vertices a and d. According to Definition 3-14, we can write

$$\tilde{W}_{ad} = \sum_{S \in \boldsymbol{S}(ad|bc)} \tilde{U}_H \tilde{U}_{H'} \tag{21-22}$$

$$\tilde{W}_{ad}^{(ad)} = \sum_{S \in \boldsymbol{S}(a=d|bc)} \tilde{U}_H^{(ad)} \tilde{U}_{H'} \tag{21-23}$$

where $\boldsymbol{S}(a = d \mid bc) \equiv \boldsymbol{S}_{G^{(ad)}}(ad \mid bc) \supset \boldsymbol{S}(ad \mid bc)$ (a cut-set in $G^{(ad)}$ is not necessarily a cut-set in G). Therefore, the right-hand side of (21-20) is nonnegative if

$$\tilde{U}\tilde{U}_H^{(ad)} \geqq \tilde{U}^{(ad)}\tilde{U}_H \tag{21-24}$$

for any $S \in \boldsymbol{S}(ad \mid bc)$. Thus we have only to prove that

$$U_H^{(ad)}/U_H \geqq U^{(ad)}/U \tag{21-25}$$

for any subgraph H of G. We may assume $H = G - l$ without loss of generality. Since $U^{(ad)} \equiv W^{(a|d)}$, the quantity $(U^{(ad)}/U)p^2$ is nothing but $Q \equiv \sum_l \alpha_l m_l^2 - V$ of a self-energy graph having two external vertices a and d. Hence Theorem 20-6 with $p^2 = 1$ implies

$$\frac{U_{G-l}^{(ad)}}{U_{G-l}} - \frac{U^{(ad)}}{U} = \varrho_l Y_l^2 \geqq 0 \tag{21-26}$$

if l is not a cut-line; the left-hand side of (21-26) equals 0 if l is a cut-line. Thus (21-20) is nonnegative. q.e.d.

Example 1 For the square graph shown in Fig. 21-1(a), we have

$$W_a = \alpha_1\alpha_2, \quad W_b = \alpha_2\alpha_3, \quad W_c = \alpha_1\alpha_4, \quad W_d = \alpha_3\alpha_4$$

$$W_{ab} = \alpha_1\alpha_3, \quad W_{ac} = \alpha_2\alpha_4, \quad W_{ad} = 0$$

whence

$$W_a W_d = \alpha_1\alpha_2\alpha_3\alpha_4 = W_{ab}W_{ac}$$

$$W_a W_b = \alpha_1\alpha_2^2\alpha_3 \geqq 0 = W_{ac}W_{ad}$$

Example 2 For Fig. 21-1(b),

$$W_a = \alpha_1\alpha_2(\alpha_4 + \alpha_5 + \alpha_6)$$

$$W_d = \alpha_5\alpha_6(\alpha_1 + \alpha_2 + \alpha_3 + \alpha_4) + \alpha_3\alpha_4\alpha_5$$

$$W_{ab} = \alpha_1\alpha_3(\alpha_4 + \alpha_5 + \alpha_6) + \alpha_1\alpha_4\alpha_6$$

$$W_{ac} = \alpha_2\alpha_4\alpha_5$$

whence

$$W_a W_d \geqq W_{ab} W_{ac}$$

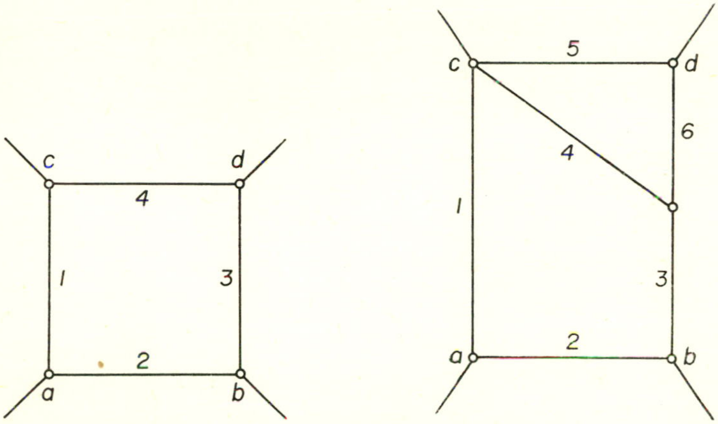

Fig. 21-1 (a) & (b) Examples for Conjecture 21-4.

In the proof of Theorem 21-3, we have used an *algebraic* inequality at the last step (21-26), but the above examples suggest that any term contained in $W_{ab}W_{ac}$ is contained also in W_aW_d. Thus we are led to the following conjecture.

CONJECTURE **21-4** [12]† When we write

$$W_a W_d = \sum_r a_r \prod_{l \in |G|} \alpha_l^{r_l}$$
$$W_{ab} W_{ac} = \sum_r b_r \prod_{l \in |G|} \alpha_l^{r_l} \qquad (21\text{-}27)$$

where $\boldsymbol{r}$ is a sequence of r_l $(l \in |G|)$, $r_l = 0$ or 1 or 2 with

$$\sum_{l \in |G|} r_l = 2(\mu + 1) \qquad (21\text{-}28)$$

† Fairlie [32] proposed a proof of Conjecture 21-4, but unfortunately it was incomplete.

and a_r and b_r are nonnegative integers, we shall always have

$$a_r \geqq b_r \quad \text{for any } \boldsymbol{r} \tag{21-29}$$

We note that a_r and b_r are *not* restricted to 0 and 1; for example, in Example 2, $a_r = 2$ for $r = (1, 1, 1, 1, 1, 1)$. Evidently, Conjecture 21-4 is more precise than Theorem 21-3. It is an interesting graph-theoretical problem to prove Conjecture 21-4.

By means of mathematical induction with respect to N, Conjecture 21-4 is reduced to the proposition that

$$a_r \geqq b_r \quad \text{for} \quad \boldsymbol{r} = (1, 1, \ldots, 1) \tag{21-30}$$

only for all graphs such that $N = 2(\mu + 1)$. This is because if there exists l such that $r_l = 0$ then the proof of (21-29) reduces to that of the similar inequality in G/l, and if there exists l such that $r_l = 2$ then it reduces to that for $G - l$ by applying $(\partial/\partial\alpha_l)^2$ to (21-27). Thus one has only to count the number of the ways of dividing $|G|$ into two 2-trees of certain types. The inequality (21-30) was explicitly checked in this way for $\mu = 1, 2, 3$ [12].† Thus Conjecture 21-4 is true at least for $N \leqq 9$.

† Unfortunately, Table I of Ref. [12] contains some errors. First row, column g : 3 instead of 2; second row, column z : 3 instead of 1; third row, column k : 0 instead of 1.

§ 22 SUPPORT PROPERTIES IN PARTICULAR MODELS

In this section, we summarize the best known results on the support properties of PTIR for Ψ_n^r and Ψ_n^r.

22-1 Vertex Function

Since in this case we cannot put all external momenta on the mass shells, the euclidean majorization method is better than the noneuclidean one. Since Ψ_3^0 is obtained as a special case $m = 1$ of Ψ_3^1, we consider only the latter, namely, the case of the nucleon vertex function.

The euclidean majorization method [33] yields the result that any graph of Ψ_3^1 is majorized by the triangle graph shown in Fig. 22-1 in the *euclidean region* of p. From it, one can show [34] that $\varphi_3(z_a, z_b, z_c, \varkappa)$ of (17-16) vanishes unless

$$\begin{aligned}
&\varkappa \geqq (z_a z_b + z_b z_c + z_c z_a)(z_a^{-1} + z_b^{-1} + m^2 z_c^{-1}) \\
&\qquad\qquad \text{for} \quad |z_a^{-1} - z_b^{-1}| \leqq m z_c^{-1} \leqq z_a^{-1} + z_b^{-1} \\
&\varkappa \geqq (m+1)^2 (z_a + z_b) \quad \text{for} \quad m z_c^{-1} > z_a^{-1} + z_b^{-1} \qquad\qquad (22\text{-}1) \\
&\varkappa \geqq (m+1)^2 z_a + (m-1)^2 z_b + 4 z_c \quad \text{for} \quad z_b^{-1} > z_a^{-1} + m z_c^{-1} \\
&\varkappa \geqq (m-1)^2 z_a + (m+1)^2 z_b + 4 z_c \quad \text{for} \quad z_a^{-1} > z_b^{-1} + m z_c^{-1}
\end{aligned}$$

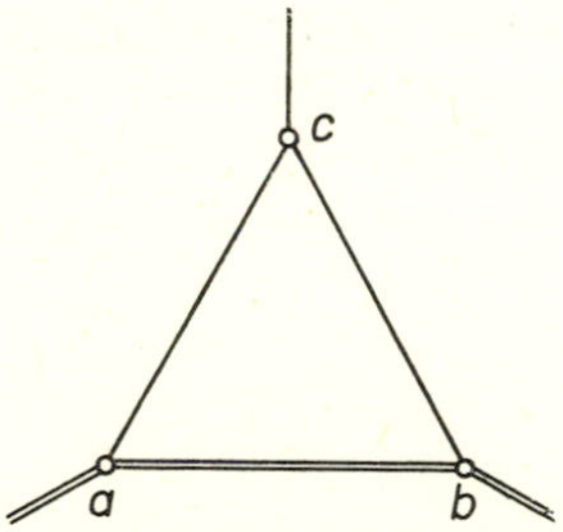

Fig. 22-1 The graph of Ω_3^1. Double lines stand for nucleon lines.

The bounding curve of the first case of (22-1) is realized in the triangle graph (Fig. 22-1). Support properties in the other cases, which arise from the contributions from outside the euclidean region is nothing but (19-18) with $M_a = M_b = m + 1$ and $M_c = 2$.

22-2 Scattering Amplitude

We consider the support properties of (17-22).

(a) Equal-mass-particle scattering (Ψ_4^0). From (21-16) we have

$$\varrho_{st}(z,\gamma) = \varrho_{tu}(z,\gamma) = \varrho_{us}(z,\gamma) = 0$$
$$\text{unless} \quad \gamma \geqq 4 + 4[z(1-z)]^{\frac{1}{2}} \tag{22-2}$$

This inequality gives us the exact shape of the support because (22-2) is realized in the square graph.

(b) Nucleon-nucleon scattering (Ψ_4^2). As above, the exact shapes of the supports are found to be

$$\varrho_{st}(z,\gamma) = 0 \quad \text{unless} \quad \gamma \geqq 4m^2z + 4(1-z) + 4[z(1-z)]^{\frac{1}{2}}$$
$$\varrho_{tu}(z,\gamma) = 0 \quad \text{unless} \quad \gamma \geqq \min[4z + 4m^2(1-z),$$
$$4m^2z + 4(1-z)] + 4[z(1-z)]^{\frac{1}{2}}$$
$$\varrho_{us}(z,\gamma) = 0 \quad \text{unless} \quad \gamma \geqq 4z + 4m^2(1-z) + 4[z(1-z)]^{\frac{1}{2}} \tag{22-3}$$

As remarked in Subsection 20-3, the unclassified majorization for ϱ_{tu} yields only $\gamma \geqq 4 + 4[z(1-z)]^{\frac{1}{2}}$. We here present two proofs of the support of ϱ_{tu} in (22-3).

Proof No. 1 [17] Since we should consider the case $s_v = m^2 + 1$ ($v = a, b, c, d$), we have only to show that $V_G \geqq 0$ for all $G \in \Psi_4^2$ for *either* $\{s = 0, t = 4, u = 4m^2\}$ *or* $\{s = 0, t = 4m^2, u = 4\}$. We choose the former set of values if two baryon paths $\boldsymbol{P}_1$ and $\boldsymbol{P}_2$ are such that $\boldsymbol{P}_1 \in \boldsymbol{P}(ac)$ and $\boldsymbol{P}_2 \in \boldsymbol{P}(bd)$; the latter set is chosen if $\boldsymbol{P}_1 \in \boldsymbol{P}(ad)$ and $\boldsymbol{P}_2 \in \boldsymbol{P}(bc)$. Because of symmetry, it is sufficient to consider the former case alone. Since $s_v = (s + t + u)/4$, we can write

$$Q \equiv \sum_l \alpha_l m_l^2 - V = \tfrac{1}{4}(Q^{(1)}s + Q^{(2)}t + Q^{(3)}u) \tag{22-4}$$

On substituting $s = 0$, $t = 4$, and $u = 4m^2$ in (22-4) and using the baryon-number conservation law

$$\sum_l \alpha_l m_l^2 \geqq \sum_l \alpha_l + \sum_{l \in P_1 \cup P_2} \alpha_l(m^2 - 1) \tag{22-5}$$

we obtain

$$V \geqq \Big[\sum_l \alpha_l - Q^{(2)} - Q^{(3)}\Big] + \Big(\sum_{l \in P_1 \cup P_2} \alpha_l - Q^{(3)}\Big)(m^2 - 1) \tag{22-6}$$

The quantity in the square bracket is just the V function for the equal-mass Feynman graph, whence it is nonnegative. To prove

$$\sum_{l \in P_1 \cup P_2} \alpha_l \geqq Q^{(3)} \tag{22-7}$$

we note that

$$Q^{(3)}_{G-(P_1 \cup P_2)^*} = \sum_{l \in P_1 \cup P_2} \alpha_l \tag{22-8}$$

where $(P_1 \cup P_2)^* \equiv |G| - P_1 \cup P_2$, provided that P_1 and P_2 pass through no common vertex [see Rule (A 3)]. Since the point $\{s_a = s_b = s_c = s_d = 1, s = t = 0, u = 4\}$ belongs to the euclidean region, Theorem 20-6 implies $Q^{(3)}_{G-(P_1 \cup P_2)^*} \geqq Q^{(3)}$. q.e.d.

Proof No. 2 [11] We employ the path method. Let $P_1 \in \boldsymbol{P}(ac)$ and $P_2 \in \boldsymbol{P}(bd)$ be two baryon paths as before. Because of Theorem 20-2 and the existence of P_1 and P_2, we can find $q_l\,(l \in |G|)$ satisfying the conditions of Theorem 19-1 if

$$\begin{aligned} p_a &= -mk_1 - k_2, \quad p_b = mk_1 + k_2 \\ p_c &= mk_1 - k_2, \quad p_d = -mk_1 + k_2 \end{aligned} \tag{22-9}$$

where $k_1^2 = k_2^2 = 1$ and $k_1 k_2 = 0$. Since (22-9) implies that

$$s_\nu = p_\nu^2 = m^2 + 1 \qquad (\nu = a, b, c, d)$$

$$s = (p_a + p_b)^2 = 0, \quad t = (p_a + p_c)^2 = 4, \quad u = (p_a + p_d)^2 = 4m^2 \tag{22-10}$$

we obtain $V \geqq 0$ for (22-10). q.e.d.

(c) Pion-pion scattering (Ψ_4^0).

If one takes account of the parity conservation law, the exact shape of the supports is found to be

$$\varrho_{st}(z, \gamma) = \varrho_{tu}(z, \gamma) = \varrho_{us}(z, \gamma) = 0$$

unless $\gamma \geqq \min\,[16z + 4(1 - z), \quad 4z + 16(1 - z)] + 16[z(1 - z)]^{\frac{1}{2}}$ (22-11)

instead of (22-2).

Proof [12] It is convenient to employ the path method. By means of Rules (A 1) and (A 2), we have only to consider strongly connected, equal-mass Feynman graphs G such that every vertex is of even F-degree. We assume that all external vertices are distinct; otherwise G is obtained as a reduced graph of another. Then, since three or more internal lines are incident with every external vertex, there exist six disjoint paths between external vertices. Topologically, there are only three possible situations A, B, C of those paths shown in Fig. 22-2. Since G is connected, however, Situation C has to be of the form D, whence it is a special case of Situation A. Since Situation A is not symmetric with respect to the external vertices, there

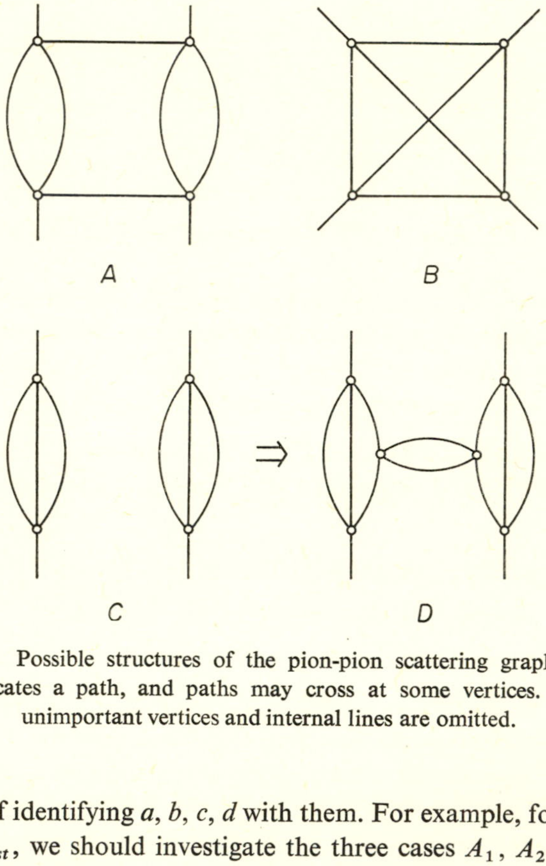

Fig. 22-2 Possible structures of the pion-pion scattering graphs. Each line indicates a path, and paths may cross at some vertices. Possible unimportant vertices and internal lines are omitted.

are six ways of identifying a, b, c, d with them. For example, for the support property of ϱ_{st}, we should investigate the three cases A_1, A_2, B shown in Fig. 22-3. We can find q_l $(l \in |G|)$ satisfying the conditions of Theorem 19-1 is shown in Fig. 22-3 if

$$p_a = -2k_1 - k_2, \quad p_b = -2k_1 + k_2$$
$$p_c = 2k_1 - k_2, \quad p_d = 2k_1 + k_2 \tag{22-12}$$

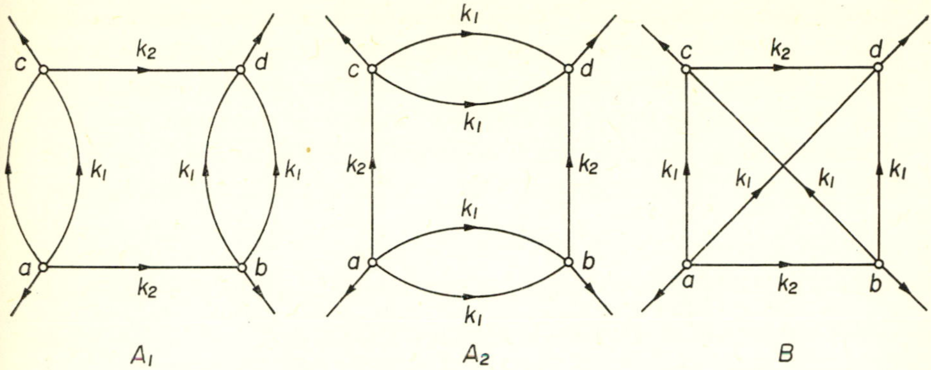

Fig. 22-3 The way of applying Theorem 19-1 to the pion-pion scattering.

for A_1 and B and if

$$\begin{aligned} p_a &= -2k_1 - k_2, \quad p_b = 2k_1 - k_2 \\ p_c &= -2k_1 + k_2, \quad p_d = 2k_1 + k_2 \end{aligned} \tag{22-13}$$

for A_2, where $k_1^2 = k_2^2 = 1$ and $k_1 k_2 = 0$. Since (22-12) and (22-13) imply $\{s_\nu = 5\,(\nu = a, b, c, d), s = 16, t = 4\}$ for A_1 and B and $\{s_\nu = 5\,(\nu = a, b, c, d), s = 4, t = 16\}$ for A_2, respectively, we obtain (22-11). q.e.d.

(d) Pion-nucleon scattering (Ψ_4^1 and Ψ_4^1). The noneuclidean majorization method yields the following results [17]. For Ψ_4^1 (b and d are of pions), one obtains that

$$\begin{aligned} &\varrho_{st}(z, \gamma) = 0 \quad \text{unless} \quad \gamma \geqq (m + 1)^2 z + 4(1 - z) + \varkappa[z(1 - z)]^{\frac{1}{2}} \\ &\varrho_{tu}(z, \gamma) = 0 \quad \text{unless} \quad \gamma \geqq 4z + (m + 1)^2 (1 - z) + \varkappa[z(1 - z)]^{\frac{1}{2}} \\ &\varrho_{us}(z, \gamma) = 0 \quad \text{unless} \quad \gamma \geqq (m + 1)^2 + \lambda[z(1 - z)]^{\frac{1}{2}} \end{aligned} \tag{22-14}$$

with $\varkappa = \lambda = 4$. For Ψ_4^1, one obtains (22-14) with $\varkappa = 8$ and $\lambda = 10$ under some plausible assumptions for certain low-order graphs. Furthermore, if one employs also the euclidean majorization method, one can improve (22-14) except for neighborhoods of $z = 0, \frac{1}{2}, 1$. Contrary to the above three cases, one does not yet succeed in proving the exact shapes of the supports.

(e) Mandelstam bounding curves.

For Ψ_4^0, $\tilde{\psi}_4^0$, and Ψ_4^2, we have obtained the exact shapes of the supports. In the Mandelstam representation (18-30), the boundaries of the supports of the double spectral functions $\sigma_{st}(s', t')$, $\sigma_{tu}(t', u')$, and $\sigma_{tu}(u', s')$ are called the

Mandelstam bounding curves. Their equations were conjectured† on the basis of the unitarity of the S-matrix in the elastic region [36]. We show that the Mandelstam bounding curves can be derived for the equal-mass, pion-pion, and nucleon-nucleon scatterings if we assume the validity of the Mandelstam representation.

Suppose

$$\varrho_{st}(z, \gamma) = 0 \quad \text{unless} \quad \gamma \geqq c_s z + c_t(1 - z) + c[z(1 - z)]^{\frac{1}{2}} \tag{22-15}$$

As shown in Subsection 18-2, $\varrho_{st}(z, \gamma)$ is uniquely related with $\sigma_{st}(s', t')$ through (18-31). To be consistent with (22-15), we should have that‡

$$\sigma_{st}(s', t') = 0 \quad \text{unless} \quad z(s' - c_s) + (1 - z)(t' - c_t) \geqq c[z(1 - z)]^{\frac{1}{2}} \tag{22-16}$$

for any $0 < z < 1$. Since the envelope of a family of those straight lines in the (s', t') plane is the right branch of a hyperbola

$$(s' - c_s)(t' - c_t) = c^2/4 \tag{22-17}$$

we see $\sigma_{st}(s', t') = 0$ unless the point (s', t') lies in the upper region bounded by it. For the above-mentioned three cases, (22-17) yields the conjectured Mandelstam bounding curves precisely.

22-3 Proof of Dispersion Relations

As an immediate consequence of PTIR, we can prove dispersion relations for t fixed.

Suppose that the following support properties are proven already:

$$\begin{aligned} \varrho_{st}(z, \gamma) &= 0 \quad \text{unless} \quad \gamma \geqq c_s z + c_t(1 - z) \\ \varrho_{tu}(z, \gamma) &= 0 \quad \text{unless} \quad \gamma \geqq c_t z + c_u(1 - z) \\ \varrho_{us}(z, \gamma) &= 0 \quad \text{unless} \quad \gamma \geqq c_u z + c_s(1 - z) + c \min(z, 1 - z) \end{aligned} \tag{22-18}$$

Since t is fixed to a real value, we substitute in (17-22)

$$\begin{aligned} s' &= z^{-1}[\gamma - t(1 - z)] \quad \text{in} \quad \varrho_{st} \\ u' &= (1 - z)^{-1}(\gamma - tz) \quad \text{in} \quad \varrho_{tu} \\ s' &= (1 - 2z)^{-1}[\gamma - (c_0 - t)z] \quad \text{for} \quad z < \tfrac{1}{2} \quad \text{in} \quad \varrho_{us} \\ u' &= (2z - 1)^{-1}[\gamma - (c_0 - t)(1 - z)] \quad \text{for} \quad z > \tfrac{1}{2} \quad \text{in} \quad \varrho_{us} \end{aligned} \tag{22-19}$$

† Recently, Martin [35] has proven the Mandelstam bounding curve for the pion-pion scattering except for $8 < s < 32$ in the axiomatic field theory.

‡ Rigorously speaking, we need reasoning similar to the proof of Theorem 18-3.

with $c_0 \equiv \sum_{\nu=a}^{d} m_\nu^2$. Then the dispersion relation

$$f_4(s) = \int_{c_s}^{\infty} ds' \frac{\alpha(s')}{s' - s} + \int_{c_u}^{\infty} du' \frac{\beta(u')}{u' - u} \tag{22-20}$$

follows if

$$-c_t \leqq -t \leqq c_s + c_u + c - c_0 \tag{22-21}$$

where $\alpha(s')$ and $\beta(u')$ are certain distributions. It is customary to denote the maximum possible value of $-t$ by $4\Delta_{\max}^2$, that is,

$$\Delta_{\max}^2 \equiv (c_s + c_u + c - c_0)/4 \tag{22-22}$$

Then the values of $\Delta_{\max}^2$ obtained from the results in Subsection 22-2 are as follows:

$\Delta_{\max}^2 = 2$ for equal-mass particle sacttering,

$\Delta_{\max}^2 = 2$ for nucleon-nucleon scattering,

$\Delta_{\max}^2 = 8$ for pion-pion scattering,

$\Delta_{\max}^2 = m + \frac{1}{4}\lambda$ for pion-nucleon scattering.

Finally, we note that PTIR can also be used for the proof of the partial-wave dispersion relation [37, 38].

22-4 One-Particle Production Amplitude

As mentioned in Subsection 20-3, Ψ_5^r reduces to Ω_5^r, which consists of the single-loop graphs alone. They do not have the leading singularity on the mass shell. Relevant singularities are those of the reduced graphs R of $n_R = 2$ and $n_R = 3$. Let a be the external vertex which is dealt with asymmetrically. Then we can show [17] that the weight distributions of (17-32) with $n = 5$ vanish unless

(a) $\Psi_5^0 : \gamma \geqq 7/3$,

(b) $\Psi_5^1 : \gamma \geqq 7/3$ if a nucleon line is incident with a,

$\gamma \geqq 2 + \sqrt{3}$ if a pion line is incident with a,

(c) $\Psi_5^1 : \gamma \geqq 4$ if a pion line is incident with a,

(d) $\Psi_5^2 : \gamma \geqq 4$.

References

(1) G. C. Wick, *Phys. Rev.* **96**, 1124 (1954).
(2) R. E. Cutkosky, *Phys. Rev.* **96**, 1135 (1954).
(3) Y. Nambu, *Phys. Rev.* **98**, 803 (1955).
(4) Y. Nambu, *Phys. Rev.* **100**, 394 (1955).
(5) S. Mandelstam, *Phys. Rev.* **112**, 1344 (1958).
(6) G. Wanders, *Helv. Phys. Acta* **30**, 417 (1957).
(7) V. Ya. Faĭnberg, *J. E. T. P.* **36**, 1503 (1959), [translation: *Sov. Phys. J. E. T. P.* **9**, 1066 (1959)].
(8) S. Deser, W. Gilbert, and E. C. G. Sudarshan, *Phys. Rev.* **115**, 731 (1959).
(9) M. Ida, *Progr. Theoret. Phys.* **23**, 1151 (1960).
(10) N. Nakanishi, *Progr. Theoret. Phys. Suppl.* **18**, 1 (1961); Errata, *Progr. Theoret. Phys.* **26**, 806 (1961).
(11) N. Nakanishi, *Progr. Theoret. Phys.* **26**, 337 (1961); Errata, *ibid.* **28**, 406 (1962).
(12) N. Nakanishi, *Progr. Theoret. Phys.* **26**, 927 (1961).
(13) N. Nakanishi, *J. Math. Phys.* **3**, 1139 (1962).
(14) N. Nakanishi, *Phys. Rev.* **127**, 1380 (1962).
(15) N. Nakanishi, *J. Math. Phys.* **4**, 1385 (1963).
(16) N. Nakanishi, *J. Math. Phys.* **5**, 1458 (1964).
(17) J. B. Boyling, *Ann. Phys.* **28**, 435 (1964).
(18) H. Umezawa and S. Kamefuchi, *Progr. Theoret. Phys.* **6**, 543 (1951).
(19) G. Källén, *Helv. Phys. Acta* **25**, 417 (1952).
(20) H. Lehmann, *Nuovo Cim.* **11**, 342 (1954).
(21) M. Gell-Mann and F. E. Low, *Phys. Rev.* **95**, 1300 (1954).
(22) N. Nakanishi, *J. Math. Phys.* **4**, 1539 (1963).
(23) A. Z. Patashinskiĭ, *J. E. T. P.* **43**, 1371 (1962), [translation: *Sov. Phys. J. E. T. P.* **16**, 973 (1963)].
(24) J. Bros and V. Glaser, *L'envéloppe d'holomorphie de l'union de deux polycercles* (1961).
(25) K. Symanzik, *Progr. Theoret. Phys.* **20**, 690 (1958).
(26) N. Nakanishi, *Progr. Theoret. Phys.* **21**, 135 (1959).
(27) H. J. Bremermann, R. Oehme, and J . G. Taylor, *Phys. Rev.* **109**, 2178 (1958).
(28) Y. Nambu, *Nuovo Cim.* **9**, 610 (1958).
(29) T. T. Wu, *Phys. Rev.* **123**, 678 (1961).
(30) J. B. Boyling, *Ann. Phys.* **25**, 249 (1963).
(31) J. B. Boyling, *J. Math. Phys.* **6**, 1469 (1965).
(32) D. B. Fairlie, *Proc. Camb. Phil. Soc.* **59**, 157 (1963).
(33) A. A. Logunov, I. T. Todorov, and N. A. Chernikov, *J. E. T. P.* **42**, 1285 (1962), [translation: *Sov. Phys. J. E. T. P.* **15**, 891 (1962)].
(34) Liu I-Ch'en and I. T. Todorov, *Doklady* **148**, 806 (1963), [translation: *Sov. Phys. Dokl.* **8**, 157 (1963)].
(35) A. Martin, *Nuovo Cim.* **44**, 1219 (1966).
(36) S. Mandelstam, *Phys. Rev.* **115**, 1752 (1959).
(37) K. Yamamoto, *Progr. Theoret. Phys.* **26**, 1014 (1961).
(38) N. Nakanishi, *Phys. Rev.* **126**, 1225 (1962).

CHAPTER 5

Miscellaneous Topics

In this chapter, we discuss some topics concerning Feynman graphs which are not related to the analyticity of the Feynman integral.

First, we count the number of the Feynman graphs having M vertices which appear under a given interaction Lagrangian. Here all vertices are regarded as distinguishable. It is extremely difficult to count the number of topologically distinct graphs, and we are not concerned with this problem.

Second, by making use of the above result, we prove the divergence of the unrenormalized perturbation series for the case of spinless particles alone.

The third topic is a graph-theoretical problem concerning Feynman graphs. We investigate how the topological structure of Feynman graphs is restricted if certain types of cut-sets are absent.

Finally, we briefly discuss the asymptotic behavior of the Feynman integral when one of the invariant squares is very large.

§ 23 NUMBER OF FEYNMAN GRAPHS

Given an interaction Lagrangian density $\mathscr{L}_{\text{int}}(x)$, the S-matrix is given by $S = \tilde{S}/\langle 0 \,|\tilde{S}|\, 0\rangle$ with (6-13). Any S-matrix element or (extended) transition amplitude is expressed in a perturbation series. Its M-th order term is a sum of the Feynman integrals corresponding to the Feynman graphs having M vertices, each of which is admissible under $\mathscr{L}_{\text{int}}(x)$ in the sense explained in Subsection 6-1. The disconnected Feynman graphs involving vacuum polarization parts are removed by the factor $\langle 0\,|\tilde{S}|0\rangle$, but, for the moment, we include them.

Let G be a Feynman graph having M vertices. As mentioned briefly in Subsection 7-3, we can assign the M positions $x_1, \ldots, x_M$ to the M vertices generally in $M!$ different ways; this number $M!$ just cancels the denominator of the coefficient in (6-13). For certain graphs, which are invariant under some permutations of vertices, however, such a cancellation is not complete so that some fractional factor remains. To avoid this complication, in this section, we regard all vertices as distinguishable, that is, two topologically identical Feynman graphs with the same particle assignment are regarded as distinct if the names of the corresponding vertices are not identical. Furthermore, for the moment, we assume that $\mathscr{L}_{\text{int}}(x)$ is *linear* with respect to each distinct field operator.

In an exceptional case, the S-matrix element may admit the presence of external particles passing without interaction, that is, we may have some "pairs of external lines" which are incident with *no* vertex. The contribution from such a case is trivially factored out, and the S-matrix element reduces to the one having a smaller number of external lines. We exclude this trivial case in accord with our definition of the Feynman graph stated in Subsection 1-4.

All fermions and some bosons are distinct from their antiparticles. For such particles,† the corresponding field operators $\psi(x)$ are different from their conjugate fields $\bar{\psi}(x)$, which are essentially the hermitian conjugates of $\psi(x)$. The other bosons are not distinguishable from their own antiparticles, that is, the corresponding fields are hermitian. To indicate this, those boson fields are denoted by $\varphi(x)$ [or $\varphi_j(x)$] in this section.

We may specify the essential structure of $\mathscr{L}_{\text{int}}(x)$ by writing field operators only symbolically. We call $\mathscr{L}_{\text{int}} \sim \bar{\psi}\psi\varphi$ the *Yukawa coupling*,

† For brevity, we call them fermions only in this section.

$\mathscr{L}_{\text{int}} \sim (\bar{\psi}_1\psi_2 + \bar{\psi}_2\psi_1)\varphi$ with $\psi_1 \neq \psi_2$ the *modified Yukawa coupling*, and $\mathscr{L}_{\text{int}} \sim \prod_{j=1}^{k} \varphi_j$ the *multilinear boson coupling*. We count the number of Feynman graphs by "drawing" lines between M given distinct vertices. We note that we can *independently* draw the lines of different-type particles to construct Feynman graphs [1].

We first consider the $\bar{\psi}\psi$ part of the Yukawa coupling. All lines corresponding to fermions have an intrinsic orientation in such a way that one outgoing and one ingoing lines are incident with every vertex. Let r be the number of outgoing fermion lines, which is equal to that of ingoing ones. The number of the ways of choosing r outgoing external vertices is $\binom{M}{r}$. From the remaining $M - r$ vertices, $M - r$ fermion internal lines are outgoing; their ingoing end points have to be selected from M vertices (then the ingoing external vertices are uniquely determined). The number of the ways of doing this is $\binom{M}{M-r}(M-r)!$. Thus the number of the ways of drawing fermion lines for the $\bar{\psi}\psi$ part is given by

$$N_F(M, r) = \binom{M}{r}\binom{M}{M-r}(M-r)! = \frac{(M!)^2}{(r!)^2\,(M-r)!} \tag{23-1}$$

Next, we consider the $\bar{\psi}_1\psi_2 + \bar{\psi}_2\psi_1$ part. Let r_i and $\bar{r}_i\,(i = 1, 2)$ be the numbers of the outgoing and ingoing external lines, respectively, of the ψ_i particle. Of course $r_1 + r_2 = \bar{r}_1 + \bar{r}_2$. Each vertex is classified into either of two classes 1 and 2 according to the name of the outgoing particle incident with it. Let $M_i\,(i = 1, 2)$ be the number of Class i; then we have $M = M_1 + M_2$, $M_1 = r_1 + (M_2 - \bar{r}_1)$, and $M_2 = r_2 + (M_1 - \bar{r}_2)$. Hence $M + r_i - \bar{r}_i$ has to be even and

$$M_i = (M + r_i - \bar{r}_i)/2 \qquad (i = 1, 2) \tag{23-2}$$

The number of the ways of dividing M vertices into two classes is $\binom{M}{M_1}$. In Class 1, r_1 vertices are assigned to the outgoing ψ_1 external lines, and the ψ_1 internal lines outgoing from the remaining $M_1 - r_1 = M_2 - \bar{r}_1$ vertices choose their ingoing end points from M_2 vertices of Class 2. Thus the number of the ways of drawing the ψ_1 lines is

$$\binom{M_1}{r_1}\binom{M_2}{M_2 - \bar{r}_1}(M_2 - \bar{r}_1)! \tag{23-3}$$

We have a similar formula for the ψ_2 lines. Thus, the number for the $\bar{\psi}_1\psi_2 + \bar{\psi}_2\psi_1$ part is given by

$$N'_F(M, r_1, r_2, \bar{r}_1, \bar{r}_2) = \frac{M!\,[(M + r_1 - \bar{r}_1)/2]!\,[(M + r_2 - \bar{r}_2)/2]!}{r_1!\,\bar{r}_1!\,r_2!\,\bar{r}_2!\,[(M - r_1 - \bar{r}_1)/2]!\,[(M - r_2 - \bar{r}_2)/2]!} \tag{23-4}$$

The number of the ways of drawing the boson (φ_j) lines is counted as follows. Let n_j be the number of the external lines of the φ_j particles. Since we assume that only one φ_j line is incident with every vertex, the number of the ways of drawing the φ_j lines is equal to that of classifying the $M - n_j$ vertices into pairs, that is, it is given by

$$N_B(M, n_j) = \binom{M}{n_j}(M - n_j - 1)!! \tag{23-5}$$

where $M - n_j$ has to be even, and $(2m - 1)!! \equiv \prod_{k=1}^{m}(2k - 1)$.

On combining the above results, we can find the number of Feynman graphs (having M vertices and n external lines) generated by the above-mentioned interaction Lagrangians. We have

$$N_F(M, r)\, N_B(M, n - 2r) \quad \text{for the Yukawa coupling,} \tag{23-6}$$

$$N'_F(M, r_1, r_2, \bar{r}_1, \bar{r}_2)\, N_B(M, n - r_1 - r_2 - \bar{r}_1 - \bar{r}_2) \quad \text{for the modified Yukawa coupling,} \tag{23-7}$$

$$\prod_{j=1}^{k} N_B\,(M, n_j] \quad \text{with} \quad \sum_{j=1}^{k} n_j = n$$

for the multilinear boson coupling. (23-8)

In (23-8), all φ_j fields are distinct. If they are all identical, $\prod_{j=1}^{k}\varphi_j$ should be replaced by $\varphi^k/k!$. The number of the graphs in the φ^k coupling case cannot be obtained from (23-8), because the graphs which are mutually transmuted by permuting $\varphi_1, \ldots, \varphi_k$ (totally or partially) are not distinguishable in the φ^k coupling case. Furthermore, in this case, there are certain special graphs which cannot be generated by the $\prod_{j=1}^{k}\varphi_j$ coupling.† Therefore,

† For example, consider self-energy graphs with $M = 3$, $N = 5$, and $k = 4$.

we have to content ourselves with obtaining upper and lower bounds on the number of the graphs.†

Let $N_k(M, n)$ be the number of the Feynman graphs having M vertices and n external lines in the φ^k coupling case. Of course, $kM - n = 2N$ has to be even. To find an upper bound on $N_k(M, n)$, for a moment we suppose that all internal lines are distinguishable. There are $\frac{1}{2}M(M + 1)$ ways of choosing two end points of any internal line and M ways of choosing one end point of any external line. Hence if we neglect the fact that each vertex is of degree k, the number of the ways of drawing all lines is $[\frac{1}{2}M(M + 1)]^N M^n$. Since the internal lines should not be distinguished, we have

$$N_k(M, n) < M^{2N+n}/N! \tag{23-9}$$

with $N = (kM - n)/2$. By means of Stirling's formula, the right-hand side of (23-9) asymptotically behaves like

$$M^{kM/2} \tag{23-10}$$

apart from a factor of order $(\text{const})^M$.

Next, we consider a lower bound. In this case, it is preferable to exclude the Feynman graphs which include vacuum polarization parts. We here consider only the Feynman graphs which have a Hamilton circuit [2], where a *Hamilton circuit* is a circuit which passes through all vertices. Those graphs are of course strongly connected, but strongly connected graphs do not necessarily have a Hamilton circuit. We first consider the $\varphi_0^2 \prod_{j=1}^{k-2} \varphi_j$ coupling. Let $N_k'(M, n)$ be the number of the Feynman graphs which have M vertices, n external lines, and a Hamilton circuit composed by the φ_0 lines alone. Then we evidently have

$$N_k'(M, n) = \tfrac{1}{2}(M - 1)! \sum_{(n_1, \ldots, n_{k-2})} \prod_{j=1}^{k-2} N_B(M, n_j) \tag{23-11}$$

where the summation goes over all partitions such that $\sum_{j=1}^{k-2} n_j = n$. When we identify all the fields $\varphi_0, \ldots, \varphi_{k-2}$, we have to take care of the fact that several graphs become indistinguishable.‡ Consider an arbitrary graph G in the φ^k coupling case. The number of Hamilton circuits in G cannot exceed $\frac{1}{2}k(k - 1)^{M-2}$ because every vertex is of degree k or less. Further-

† We also note that the Feynman integral involves some extra factors, that is, each of r parallel lines of identical particles induces a factor $1/r!$.

‡ This point is not taken into account in Ref. [2].

more, given a Hamilton circuit C of G, the number of the ways of assigning $\varphi_1, \ldots, \varphi_{k-2}$ to all the lines of $|G| - C$ and the external lines of G at random is $(k-2)^{N-M+n}$. Therefore, the number of the graphs in the $\varphi_0^2 \prod_{j=1}^{k-2} \varphi_j$ coupling case to which G corresponds cannot exceed $\frac{1}{2}k(k-1)^{M-2}(k-2)^{N-M+n}$. Thus the number $\tilde{N}_k(M, n)$ of the strongly connected Feynman graphs in the φ^k coupling case is bounded by

$$\tilde{N}_k(M, n) > N'_k(M, n)/k^{N+n-1} \tag{23-12}$$

The right-hand side of (23-12) asymptotically behaves like (23-10). Therefore, the number of the topologically different, strongly connected Feynman graphs generated by the φ^k coupling asymptotically behaves like

$$\tilde{N}_k(M, n)/M! \sim M^{(k-2)M/2} \tag{23-13}$$

apart from a factor of order $(\text{const})^M$. We can also find a lower bound of the same order on the number of the strongly connected Feynman graphs which involve no self-energy parts.

§ 24 DIVERGENCE OF THE PERTURBATION SERIES

As stated in Section 23, the number of the Feynman graphs having M vertices (in nontrivial theories) increases asymptotically like $M^{M/2}$ or more. Hence unless the contributions from them cancel with each other very strongly, the perturbation series will diverge for any nonzero value of the coupling constant. In fact, such cancellations do not occur in the theory involving bosons only. In the theory involving fermions, however, it may be possible that such cancellations happen because of the Pauli principle for fermions.† Indeed, according to Yennie and Gartenhaus [3]‡, the unrenormalized perturbation series for the Yukawa-coupling theory (namely, $\mathscr{L}_{\text{int}} \sim \bar{\psi}\psi\varphi$ where ψ is a fermion field and φ is a boson one) is convergent for all values of the coupling constant *if the space-time volume is finite* and if high momenta are cut off, and therefore the transition amplitude is an entire function of the coupling constant. The reason for this result can be explained as follows. Under the above assumptions, the number of the possible quantum-mechanical states of fermions becomes a finite constant, and therefore the contribution from the fermion modes increases like at most $(\text{const})^M$. Accordingly, we have only to count the number of the ways of drawing boson lines. Thus the M-th order term of the perturbation series asymptotically behaves like $M^{-M/2}$, and the series is absolutely bounded by an exponential function of the coupling constant squared.

Thirring and others [5, 6, 7, 8] showed that the ordinary unrenormalized perturbation series for the theory involving spinless particles alone is divergent for any nonzero value of the coupling constant. We here present a proof of this divergence based on the Feynman-parametric integral [9, 2].

In order to avoid ultraviolet divergence, we have *either* to cut off high momenta *or* to employ a two-dimensional model. We here employ the first alternative. By using the Feynman cutoff (10-29) with $\lambda_l = 1$, the cutoff Feynman integral reads

$$f_G(p, \Lambda) = (2M - 3)! \prod_l \left(\int_{m_l^2}^{\Lambda_l^2} d\varkappa_l^2 \right) \int_\Delta d^{N-1}\alpha \, \frac{\prod_l \alpha_l}{U^2(V - i\varepsilon)^{2M-2}} \tag{24-1}$$

where all $m_l (l \in |G|)$ in V are replaced by $\varkappa_l$, respectively. If p belongs to the domain

$$D_0 = \{p \mid V(p, \alpha) > 0 \quad \text{for all} \quad \alpha \in \Delta\} \tag{24-2}$$

† The Pauli principle forbids two identical fermions to be in the same state.

‡ For related work, see Simon's paper [4].

then as $\varepsilon \to +0$ (24-1) is absolutely convergent, and $f_G(p, \Lambda)$ is positive definite. Let $V_{\max}$ be the maximum value of V with $\varkappa_l^2 = \Lambda_l^2$ $(l \in |G|)$ for p fixed. Then

$$f_G(p, \Lambda) \geqq (2M-3)! \frac{\prod_l (\Lambda_l^2 - m_l^2)}{V_{\max}^{2M-2}} \int_\Delta d^{N-1}\alpha \prod_l \alpha_l / U^2 \tag{24-3}$$

Since Theorem 9-3 implies

$$U < 2^N / N^{N-M+1} \tag{24-4}$$

we have

$$f_G(p, \Lambda) > \frac{N^{2(N-M+1)}(2M-3)!}{2^{2N}(2N-1)!} \frac{\prod_l (\Lambda_l^2 - m_l^2)}{V_{\max}^{2M-2}} \tag{24-5}$$

If $\mathscr{L}_{\text{int}}$ is of degree k with respect to boson fields, we have $\frac{1}{2} \leqq N/M \leqq k/2$ for $M \geqq 2$. Hence, according to Stirling's formula, (24-5) implies asymptotically

$$f_G(p, \Lambda) > A^M \tag{24-6}$$

where A is a certain positive quantity independent of M. For simplicity of description, suppose that $\mathscr{L}_{\text{int}}$ contains only one coupling constant g. Then, according to (23-13), the perturbation series for the extended transition amplitude is of the form

$$i \sum_{M=1}^{\infty} g^M \chi(M) \tag{24-7}$$

where $\chi(M) > 0$ for $kM - n (\geqq 0)$ even, and $\chi(M)$ asymptotically increases as $M^{(k-2)M/2}$ (or more). Thus the perturbation series is divergent for any nonzero value of g.

We remark that the above proof also applies to the so-called parity-violating Yukawa coupling [10]†

$$\mathscr{L}_{\text{int}}(x) = g\, \bar{\psi}(x)\,(1 + i\gamma^{(5)})\, \psi(x)\, \varphi(x) \tag{24-8}$$

with

$$\gamma^{(5)} \equiv \gamma^{(0)}\gamma^{(1)}\gamma^{(2)}\gamma^{(3)}$$
$$\gamma^{(5)}\gamma^{(\nu)} + \gamma^{(\nu)}\gamma^{(5)} = 0 \qquad (\nu = 0, 1, 2, 3) \tag{24-9}$$
$$(\gamma^{(5)})^2 = -1$$

because the numerator of the fermion Feynman propagator reduces to a constant when sandwiched by $1 + i\gamma^{(5)}$:

$$(1 + i\gamma^{(5)})\,(m_l + \gamma k_l)\,(1 + i\gamma^{(5)}) = 2m_l(1 + i\gamma^{(5)}) \tag{24-10}$$

Since the convergence proof of Yennie and Gartenhaus applies also to (24-8), we see that the assumption of the finite space-time volume is more essential for the convergence than the Pauli principle is.

† Though (24-8) is not hermitian, this point is irrelevant to our reasoning.

Finally, we briefly mention the renormalized perturbation series. Here we are concerned with mass and coupling-constant renormalizations rather than the removal of ultraviolet divergence itself. Unrenormalized masses and coupling constants are expressed in power series of renormalized masses and coupling constants. The amounts of the changes from the former to the latter are called *renormalization constants*. The *renormalized perturbation series*, which is expressed in terms of renormalized masses and coupling constants, is formally equal to the unrenormalized perturbation series, which is expressed in terms of unrenormalized masses and coupling constants. Thus the mass renormalization implies certain rearrangement of the perturbation series, and the coupling-constant renormalization implies the change of expansion parameters. As stated in Subsection 10-2, all Feynman integrals become free from ultraviolet divergence without cutoff in the renormalizable field theories.

For definiteness, we consider the case $\mathscr{L}_{\text{int}} \simeq g\varphi^k$ with $g \neq 0$ real. Since $\chi(M) = 0$ for $kM - n$ odd, all terms of unrenormalized perturbation series have the *same sign* unless $g < 0$ and k is even. In that case, therefore, at least one of renormalization constants should be divergent if we hope that the renormalized perturbation series would be convergent. This divergence is independent of ultraviolet divergence because we are considering Feynman-cutoff quantities. This result may also be understood in the following way. It can be shown [11] that if all renormalization constants are assumed to be finite, the $g\varphi^k$ theory has no ground state (zero-energy state) except for the case in which $g < 0$ and k is even (only in this case the expectation value of the Hamiltonian is positive definite). As seen from (17-15), however, the perturbation series for the self-energy has no imaginary part for $s < 0$. This fact physically implies that the theory has a ground state if the renormalized perturbation series is meaningful. Thus, to be consistent, except in the case $g < 0$ and k even, *either* at least one of renormalization constants is divergent *or* the renormalized perturbation series is divergent.

It is extremely difficult to show convergence or divergence of the completely renormalized series, but it is easy to discuss the series for which mass renormalization alone is carried out. We have only to exclude all Feynman graphs involving self-energy parts and to replace the propagator (6-15) by

$$-i\left[\frac{1}{\tilde{m}_l^2 - k_l^2 - i\varepsilon_l} + \int_{c_l}^{\infty} ds' \frac{\varrho(s')}{s' - k_l^2 - i\varepsilon_l}\right] \tag{24-11}$$

where $\tilde{m}_l$ denotes the renormalized mass and we know from the axiomatic field theory that $c_l > \tilde{m}_l^2$ and $\varrho(s') \geqq 0$. By taking the first term of (24-11) as a lower bound, we can still make the same proof of the divergence of the series. For $k = 3$, all Feynman integrals become finite after mass renormalization, whence the mass-renormalized perturbation series can be considered without cutoff, and it is divergent.

§ 25 CUT-SET PROPERTIES OF FEYNMAN GRAPHS

The Feynman integrals corresponding to certain Feynman graphs G do not contribute to some or any weight distributions (except for the end points of the z variables) of PTIR for the transition amplitude (see Section 17) and to some or any double spectral functions of the Mandelstam representation for the scattering amplitude (see Subsection 18-2). The presence or absence of the contributions is determined by some topological properties of G, namely, by the existence or nonexistence of cut-sets of certain types. Thus we are led to a graph-theoretical problem.

Let $g = \{a, b, c, \ldots\}$ be the set of all external vertices of G as before. They are not necessarily distinct because they are supposed to be in one-to-one correspondence to the external lines of G as in Chapter 4. A cut-set $S \in \boldsymbol{S}(h \mid g - h)$ is called of *mass type* if either h or $g - h$ consists of only one external vertex, and it is called of *energy type* if both h and $g - h$ contain at least two external vertices. Of course, there exist cut-sets of energy type only when $n \geqq 4$. The structure of a connected Feynman graph G is much restricted if some sets $\boldsymbol{S}(h \mid g - h)$ of energy-type cut-sets are empty. We call this property the *cut-set property* of Feynman graphs.

In the following, we consider the case $n = 4$ only. The sets of energy-type cut-sets are $\boldsymbol{S}(ab \mid cd)$, $\boldsymbol{S}(ac \mid bd)$, and $\boldsymbol{S}(ad \mid bc)$. The number of nonempty ones among them is denoted by σ; $\sigma = 0, 1, 2, 3$. We can prove the following theorems, which appear to be rather self-evident, but their proofs are not necessarily trivial.

THEOREM **25-1** If $\sigma \geqq 2$ then all four sets of mass-type cut-sets are nonempty.

THEOREM **25-2** A necessary and sufficient condition for $\sigma = 0$ is the existence of a vertex e such that all paths† between any pair of external vertices pass through it. This vertex e is a cut-vertex if all or three external vertices are distinct.

THEOREM **25-3** A necessary and sufficient condition for $\sigma \leqq 1$, say $\boldsymbol{S}(ac \mid bd) = \boldsymbol{S}(ad \mid bc) = \emptyset$ for definiteness, is the existence of at least one vertex e such that any path between a or b and c or d passes through it. This vertex e is a cut-vertex if a, b, c, d are all distinct.

† If $P \in \boldsymbol{P}(ab)$ and if a and b are not distinct, then we understand that $P = \emptyset$ passes through $a = b$. The same applies also to Theorem 25-3.

We outline the proofs of the above theorems. We can prove Theorem 25-1 by noticing that if $S \in \mathbf{S}(ab \mid cd)$ and $S' \in \mathbf{S}(ac \mid bd)$ then there exists a cut-set $\tilde{S} \in \mathbf{S}(v \mid g - v)$ for any $v = a, b, c, d$ such that $\tilde{S} \subset S \cup S'$. For Theorems 25-2 and 25-3, only the proofs of necessity are nontrivial. The graph G under consideration will in general consist of several maximal nonseparable subgraphs. Any two of them have at most one common vertex, which is a cut-vertex. We can successively contract nonseparable subgraphs which have only one cut-vertex and at most one external noncut-vertex of G without destroying any cut-set of energy type. Let R be the reduced graph of G which cannot further be contracted in this way. For $\sigma = 0$, we can show that R is nonseparable and that three external vertices are not distinct in R. Theorem 25-2 follows from this result. It is sufficient to prove Theorem 25-3 in the case in which a, b, c, d are all distinct in R and $\mathbf{S}(ab \mid cd) \neq \emptyset$. Then, R has at least one cut-vertex e such that any paths between a or b and c or d passes through e. The existence of such a cut-vertex follows from Menger's theorem [12], which states: In a nonseparable graph, for any two distinct vertices a' and c', there exist two disjoint paths between a' and c' which pass through no common vertices except for a' and c'. If R is nonseparable, we construct a graph R' from R by adding two vertices a' and c' and four lines which link a' and a, a' and b, c' and c, and c' and d. Then by applying Menger's theorem to R', we see that R has a cut-set belonging to $\mathbf{S}(ac \mid bd)$ or $\mathbf{S}(ad \mid bc)$. This contradicts $\sigma \leqq 1$.

For $\sigma \leqq 1$, V can depend on only one of $s \equiv s_{ab}$, $t \equiv s_{ac}$, $u \equiv s_{ad}$ because at most only one of W_{ab}, W_{ac}, and $W_{ad}(W_h \equiv W^{(h|g-h)})$ is not identically zero. Thus Theorem 25-3 characterizes the Feynman graphs which contribute to no weight distributions of (17-22) except for $z = 0, 1$, that is, they contribute to only the single-spectral terms in the Mandelstam representation (18-30).

Now, we consider the problem of characterizing the Feynman graphs of $\sigma = 2$. From the above consideration, we may assume without loss of generality that G is nonseparable. Let R be the reduced graph which is obtained from G by contracting all its self-energy parts and vertex parts.† We call R the *skeleton* of G.

THEOREM **25-4** [13] If the skeleton R of a nonseparable graph G is a planar Feynman graph, then $\sigma \leqq 2$.

† For their definitions, see Subsection 1-4.

Proof If a cut-set $S \in \boldsymbol{S}(h \mid g - h)$ intersect a self-energy part or a vertex part H, we can construct a cut-set $S' \in \boldsymbol{S}(h \mid g - h)$ by replacing $S \cap |H|$ by one (or two) external line of H which is internal in G. Therefore, if $\boldsymbol{S}(h \mid g - h) = \emptyset$ in R, then $\boldsymbol{S}(h \mid g - h) = \emptyset$ in G, too. Hence we have only to show $\sigma \leqq 2$ in R. Since R is planar, we can realize it on a plane in such a way that all external vertices lie on the outermost circuit. Suppose that it passes through a, b, d, c in this order, say. Then, as is seen by a geometrical consideration, any path $P \in \boldsymbol{P}(ad)$ crosses any path $P' \in \boldsymbol{P}(bc)$, that is, they pass through a common vertex. Hence $\boldsymbol{S}(ad \mid bc) = \emptyset$. q.e.d.

For $\sigma \leqq 2$, one of W_{ab}, W_{ac}, and W_{ad} has to vanish identically. Hence Theorem 25-4 presents a sufficient condition for the Feynman graphs which contribute to a single term of (17-22) or to only one double spectral function of the Mandelstam representation.

The inverse proposition of Theorem 25-4 is not necessarily true in general. For example, the graph shown in Fig. 25-1 has a nonplanar skeleton, but $\sigma = 2$. If we restrict the vertices to have F-degree 3, however, we can exclude such counterexamples.

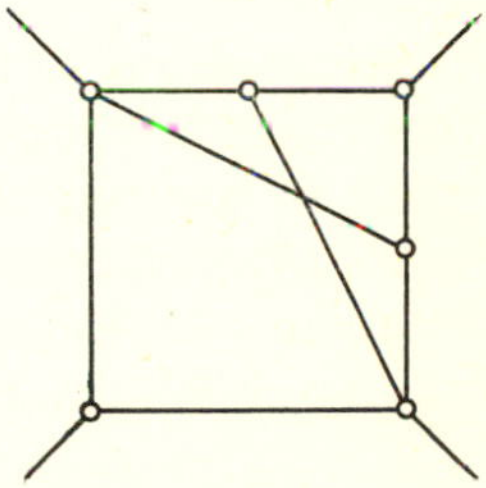

Fig. 25-1 A counterexample to the inverse proposition of Theorem 25-4.

CONJECTURE **25-5**† If in a nonseparable Feynman graph G the number of lines (including external ones) incident with every vertex is three and if $\boldsymbol{S}(ad \mid bc) = \emptyset$, then the skeleton of G is a planar Feynman graph, in which the outermost circuit passes through a, b, d, c in this order.

It is an interesting graph-theoretical problem to prove Conjecture 25-5.

† Tiktopoulos [13] proposed Conjecture 25-5 as a theorem, but unfortunately his proof was incomplete. The author is grateful to him for correspondences.

§26 HIGH-ENERGY ASYMPTOTIC BEHAVIOR OF THE FEYNMAN INTEGRAL

In connection with the Regge-pole theory,† it is interesting to consider a certain infinite sum of Feynman integrals which are generated by an integral equation, such as the Bethe-Salpeter equation, whose kernel is a product of Feynman propagators (and possibly a Feynman integral). One computes the asymptotic behavior of each term when one of invariant squares is very large. For example, consider the behavior of the scattering amplitude as $|t| \to \infty$. If one finds that the m-th order scattering amplitude, obtained by iterating the kernel m times, behaves asymptotically like

$$T_4^{(m)}(s, t) \sim \beta(s) \frac{[g(s)]^m}{m!} t^{-k} (\log t)^m \tag{26-1}$$

as $|t| \to \infty$, then one conjectures that the approximate scattering amplitude $T_4(s, t)$ will behave like

$$T_4(s, t) \equiv \sum_{m=0}^{\infty} T_4^{(m)}(s, t) \sim \beta(s)\, t^{-k+g(s)} \tag{26-2}$$

for $g(s) > 0$ sufficiently small. [The behavior like (26-2) is called the Regge behavior, which is predicted by the Regge-pole theory.] This procedure is called the *leading-term summation*. Though there is no mathematical basis for its validity, this method is useful for finding asymptotic behavior when the original integral equation cannot be solved exactly.

In this section, we consider the problem of finding the asymptotic behavior of each Feynman integral as $|t| \to \infty$, because this problem is related to graph-theoretical consideration. For simplicity, we consider only the planar Feynman graphs of the φ^3 theory [14, 15].

In the φ^3 theory, all particles are spinless, and all vertices are of F-degree 3, whence for $n = 4$ we have $3M = 2N + 4$, namely, $N = 3\mu + 1$. Then (7-69) with $\lambda = 1$ reads

$$f_G(s, t) = \mu! \prod_l \left(\int_0^\infty d\alpha_l \right) \frac{\left(\sum_l \alpha_l\right) \exp\left(-\sum_l \alpha_l\right)}{U^2(V - i\varepsilon)^{\mu+1}} \tag{26-3}$$

If the coefficient of t in V did not vanish for all values of α_l $(l \in |G|)$, the asymptotic behavior of $f_G(s, t)$ as $|t| \to \infty$ would evidently be $\sim t^{-\mu-1}$. It can

† The Regge-pole theory is a theory of discussing the analyticity in the complex angular momentum in order to predict high-energy asymptotic behavior of the transition amplitude.

vanish, however, and the leading asymptotic behavior arises from such values of α_l. If we consider a planar Feynman graph G whose outermost circuit C passes through the external vertices a, b, d, c in this order, then Theorem 25-4 implies $S(ad \mid bc) = \emptyset$ so that† $W_{ad} \equiv 0$, and therefore when V is expressed as a function of s and t, the coefficient of t in $-V$ is $\zeta_{ac} \equiv W_{ac}/U \geqq 0$. Thus it can vanish only when some of α_l ($l \in |G|$) vanish.

Let P_{ac} and P_{bd} be the path between a and c and that between b and d included in C, respectively. Any path between a vertex on P_{ac} and that on P_{bd} is called a *t-path* (see Fig. 26-1). The set of all t-paths is denoted by $\tilde{\boldsymbol{P}}_t$, and its subset consisting of all the t-paths of the minimum "length" is denoted by $\boldsymbol{P}_t$, that is,

$$\boldsymbol{P}_t = \{P \mid P \in \tilde{\boldsymbol{P}}_t, N(P) \leqq N(\tilde{P}) \quad \text{for any} \quad \tilde{P} \in \tilde{\boldsymbol{P}}_t\} \tag{26-4}$$

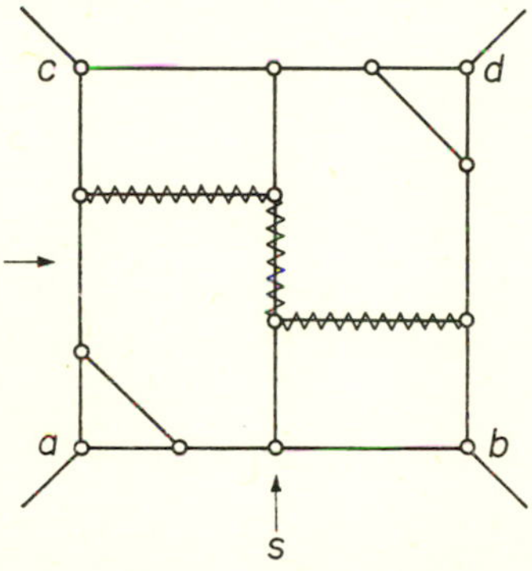

Fig. 26-1 An example of a t-path. Notched lines form a t-path.

THEOREM **26-1** In a planar Feynman graph G having four external vertices a, b, d, c as specified above, a necessary and sufficient condition for $\zeta_{ac} = 0$ but $U \neq 0$ is that $\alpha_l = 0$ for all $l \in P$, where P is a t-path.

Proof If $\alpha(P) = 0$ (i.e., $\alpha_l = 0$ for all $l \in P$) then V reduces to $V_{G/P}$ because of (9-2). Since $\boldsymbol{S}_{G/P}(ac \mid bd) = \emptyset$, we see $W_{ac} \mid_{\alpha(P)=0} = 0$, but of course $U \mid_{\alpha(P)=0} \neq 0$.

Conversely, $\zeta_{ac} = 0$ implies $W_{ac} = 0$, whence $\alpha_l(I) = 0$ for some $I \subset |G|$. Since $U \neq 0$, I includes no circuit. If there exists $S \in \boldsymbol{S}_G(ac \mid bd)$ such that

† $W_h = W^{(h|g-h)}$.

$I \cap S = \emptyset$, then the contribution, W_S, from S to W_{ac} is nonvanishing by definition. Hence for any $S \in \mathbf{S}_G(ac \mid bd)$ we have $I \cap S \neq \emptyset$, that is, $\mathbf{S}_{G/I}(ac \mid bd) = \emptyset$ on account of Theorem 2-2 (d). Since of course $\mathbf{S}_{G/I}(ad \mid bc = \emptyset$, G/I has $\sigma \leqq 1$. According to Theorem 25-3, therefore, in G/I there is a vertex e such that any path between a and c and any path between b and d pass through it. Thus I has to include a t-path in G. q.e.d.

We decompose the integration region of (26-3) into 2^N subregions by dividing the positive real axis of each α_l into two intervals $[0, \eta]$ and (η, ∞), where $\eta > 0$. Each subregion is characterized by the set, I, of all lines l such that $0 \leqq \alpha_l \leqq \eta$. Our task is to find the set I corresponding to the subregion which yields the leading asymptotic behavior of (26-3) as $|t| \to \infty$.

Let $\delta \equiv \sum_{l \in I} \alpha_l$. If I contains no circuit, Theorem 26-1 implies that as $\delta \to 0$, ζ_{ac} behaves like δ^k, where k is the maximum number of disjoint t-paths included in I. If I includes a circuit, let I' be a union of disjoint t-paths of the maximum number k included in I such that $I' \cap P_{ac} = I' \cap P_{bd} = \emptyset$ (this condition is independent of determining k). Then I' includes no circuit because every vertex is of F-degree 3. Now, as $\delta \to 0$, ζ_{ac} behaves like at best δ^{k-1} because $U = 0(\delta)$. On the other hand, as $\delta' \to 0$, where $\delta' \equiv \sum_{l \in I'} \alpha_l$, ζ_{ac} behaves like δ'^k. Thus it is sufficient to consider only the case in which I is a union of disjoint t-paths:

$$I = \bigcup_{j=1}^{k} P_j, \quad P_j \in \tilde{\mathbf{P}}_t$$
$$P_i \cap P_j = \emptyset \quad \text{for} \quad i \neq j \tag{26-5}$$

and I includes no circuit. Then the asymptotic contribution from the subregion corresponding to I is, apart from a possible logarithmic factor,

$$\int_0^\eta \frac{\delta^{m-1}\, d\delta}{(\delta^k t + c)^{\mu+1}} \sim t^{-m/k}, \qquad (c \neq 0) \tag{26-6}$$

where $m \equiv N(I)$. According to (26-6), the power of $1/t$ is equal to the average of the length $N(P_j)$ ($j = 1, \ldots, k$). Hence in order to minimize it, we should have $P_j \in \mathbf{P}_t$ for all j.

The power of $\log t$ then can be calculated as follows. Let $\delta_j \equiv \sum_{l \in P_j} \alpha_l$ ($j = 1, \ldots, k$), and $r \equiv N(P_j)$ (independent of j). By induction, we can show that

$$\prod_{j=1}^{k} \left(\int_0^{\eta} \delta_j^{r-1} \, d\delta_j \right) \left[\left(\prod_{j=1}^{k} \delta_j \right) t + c \right]^{-\mu-1} \sim t^{-r} (\log t)^{k-1} \tag{26-7}$$

Thus we obtain the following theorem.

THEOREM **26-2** [14] The Feynman integral $f_G(s, t)$ with s fixed asymptotically behaves like

$$\text{const } t^{-r} (\log t)^{k-1} \tag{26-8}$$

as $|t| \to \infty$, where $r = N(P)$ ($P \in \boldsymbol{P}_t$), and k denotes the maximum number of disjoint t-paths of $\boldsymbol{P}_t$ (their union can include no circuit).

Though Theorem 26-2 is a general rule, the constant coefficient in (26-8) is sometimes divergent. This situation can happen when G involves a self-energy part or a vertex part of certain special types, called the "singular configurations" [14]. We omit details.

For nonplanar Feynman graphs, the coefficient of t in V can vanish for none of α_l being zero. Hence the investigation of the asymptotic behavior for nonplanar graphs is more involved [13, 16].

References

(1) R. J. Riddell, Jr., *Phys. Rev.* **91**, 1243 (1953).
(2) A. Jaffe, *Commun. Math. Phys.* **1**, 127 (1965).
(3) D. R. Yennie, and S. Gartenhaus, *Nuovo Cim.* **9**, 59 (1958).
(4) B. Simon, *Nuovo Cim.* **59** A, 199 (1969).
(5) C. A. Hurst, *Proc. Roy. Soc.* A **214**, 44 (1952).
(6) W. E. Thirring, *Helv. Phys. Acta* **26**, 33 (1953).
(7) M. A. Peterman, *Helv. Phys. Acta* **26**, 291 (1953).
(8) R. Utiyama and T. Imamura, *Progr. Theoret. Phys.* **9**, 431 (1953).
(9) N. Nakanishi, *Progr. Theoret. Phys.* **17**, 401 (1957).
(10) K. Yokoyama, *Progr. Theoret. Phys.* **39**, 830 (1968).
(11) G. Baym, *Phys. Rev.* **117**, 886 (1960).
(12) C. Berge, *Théorie des Graphes et Ses Applications*, 2nd ed. (Dunod, Paris, 1963), p. 192.
(13) G. Tiktopoulos, *Phys. Rev.* **131**, 2373 (1963).
(14) G. Tiktopoulos, *Phys. Rev.* **131**, 480 (1963).
(15) M. Martinis, *J. Math. Phys.* **6**, 136 (1965).
(16) C. S. Lam, *Nuovo Cim.* **62** A, 97 (1969).

Appendix A Some Analytical Concepts

In this appendix, we explain without rigor some mathematical concepts related to distributions and analytic functions. Readers who are familiar with those subjects need not read this appendix.

A-1 Distributions

The *distribution* is a generalized concept of the function. Dirac's δ-function, its derivatives, and Cauchy's principal value are examples of distributions.

According to Schwartz [1], a distribution is a linear, continuous functional $\varphi[f]$ over test functions f. *Test functions* are functions such that they are infinitely many times differentiable and such that the asymptotic behaviors of them and their derivatives are specified. The continuity of $\varphi[f]$ means that if f_k and any derivative of f_k uniformly tend to f and the corresponding derivative of f, respectively, then $\varphi[f_k]$ tends to $\varphi[f]$.

Practically, a distribution is a generalized function such that a product of it and any test function is integrable through integrations by parts. Let $f(x)$ be any test function of n real variables $x_1, \ldots, x_n$. If

$$\varphi[f] \equiv \int \varphi(x) f(x) \, d^n x \tag{A-1}$$

is well defined, where the integration range in (A-1) is the n-dimensional euclidean space R^n (or its subregion), then $\varphi(x)$ can be regarded as a distribution practically. For example, the k-th derivative of δ-function is defined by

$$\int \delta^{(k)}(x) f(x) \, dx = (-1)^k f^{(k)}(0) \tag{A-2}$$

The *support of a function* $f(x)$ is the complement of the maximum open set in which $f(x) = 0$. The *support of a distribution* $\varphi(x)$ is the complement of the maximum open set D such that (A-1) vanishes for any $f(x)$ whose support is included in D. We denote the support of $\varphi(x)$ by supp φ. [Distinguish it from sup, which means supremum.]

Let $\{O_1, O_2, \ldots\}$ be an *open covering* of R^n, that is, let O_j (for any j) be an open set and $\bigcup_j O_j = R^n$. Then an arbitrary distribution $\varphi(x)$ can be decomposed into

$$\varphi(x) = \sum_j \varphi_j(x), \quad \operatorname{supp} \varphi_j \subset O_j \tag{A-3}$$

Thus we can speak about $\varphi(x)$ at a point x by considering its neighborhood, through the "value" of $\varphi(x)$ at x is mathematically meaningless.

As for the asymptotic behavior, in most cases, test functions are restricted to those which, together with all their derivatives, tend to zero as $|x| \to \infty$ faster than $|x|^{-\nu}$ for any $\nu > 0$. In this case, the corresponding distribution is called *tempered.* A tempered distribution has its fourier transform [2].

Any distribution which has a compact (i.e. bounded and closed) support can be represented as a certain (finite-order) derivative of a continuous function [3]. [This property can be extended to a tempered distribution.] Thus the distribution is a natural extension of the continuous function.

Let a distribution $\varphi(x)$ be a certain derivative of a continuous function $\psi(x)$. We call a point $x = a$ a *singularity* of $\varphi(x)$ if and only if $\psi(x)$ is not infinitely many times differentiable at $x = a$. Though the choice of $\psi(x)$ is not unique, this definition is not ambiguous. Dirac's δ-function $\delta(x)$ and Cauchy's principal value $\mathrm{P}\,.\,1/x$ have a singularity only at $x = 0$.

A product of two distributions $\varphi_1(x)$ and $\varphi_2(x)$ can be defined as a distribution if both of them do not have a singularity at the same point. Though $\varphi_1(x)\,\varphi_2(x)$ can be defined also in some other cases (e.g. $|x|\,\delta(x) = 0$), products of three distributions are not in general associative if two of them have a singularity at the same point. For example,

$$0 = [\delta(x) \cdot x]\,\mathrm{P}\cdot 1/x \neq \delta(x)\,[x\,\mathrm{P}\cdot 1/x] = \delta(x) \tag{A-4}$$

Let $f(x, y)$ be a real analytic function of a variable x containing parameters $y_1, \ldots, y_k$. For general values of y, all real roots $x = \alpha_1(y), \ldots, \alpha_k(y)$ of the equation $f(x, y) = 0$ will be simple, that is, they will not coincide with any root of $f'(x, y) = 0$, where $f'(x, y) \equiv (\partial/\partial x)\,f(x, y)$. Since the support of $\delta[f(x, y)]$ consists of points $x = \alpha_1(y), \ldots, \alpha_k(y)$, we have

$$\delta[f(x, y)] = \sum_{j=1}^{k} |f'(\alpha_j(y), y)|^{-1}\,\delta[x - \alpha_j(y)] \tag{A-5}$$

with the aid of

$$\delta(ax) = |a|^{-1}\,\delta(x) \quad \text{for} \quad a \neq 0, \text{real} \tag{A-6}$$

After integrating $\delta[f(x, y)]$ (multiplied by a test function of x) over x, therefore, we have a function $F(y)$ defined except for the values which yield nonsimple real roots of $f(x, y) = 0$. For those values, $F(y)$ will in general be divergent. Let $\varphi(z)$ be a distribution of one dimension whose singularities are discrete. As done in the above for $\delta(z)$, we can define $\varphi[f(x, y)]$. For example, we have

$$\mathrm{P}\cdot\frac{1}{x^2 - y^2} = \frac{1}{2y}\left(\mathrm{P}\cdot\frac{1}{x - y} - \mathrm{P}\cdot\frac{1}{x + y}\right) \tag{A-7}$$

Next, we consider a finite part of a singular function. If $f(x)$ is a test function which vanishes asymptotically faster than $|x|^{-\nu}$ for any $\nu > 0$, the integral

$$I(\lambda) \equiv \int x^{\lambda}\theta(x)\, f(x)\, dx \tag{A-8}$$

exists for $\mathrm{Re}\,\lambda > -1$, where $\theta(x) = 1$ for $x > 0$ and $\theta(x) = 0$ for $x < 0$. *Hadamard's finite part* [4] is a generalization of this integral to all values of λ not equal to negative integers. It is defined by

$$\int \mathrm{Pf}\cdot\, x^{\lambda}\theta(x)\, f(x)\, dx \equiv \lim_{\varepsilon\to 0}\left[\int_{\varepsilon}^{+\infty} x^{\lambda}f(x)\, dx + \sum_{j=0}^{k}\frac{f^{(j)}(0)}{j!}\,\frac{\varepsilon^{\lambda+j+1}}{\lambda + j + 1}\right] \tag{A-9}$$

where $\mathrm{Re}\,\lambda + k + 2 > 0$. Then it coincides with the analytic continuation of $I(\lambda)$, and therefore it satisfies

$$(d/dx)\,\mathrm{Pf}\cdot\, x^{\lambda}\theta(x) = \lambda\cdot\mathrm{Pf}\cdot\, x^{\lambda-1}\theta(x) \tag{A-10}$$

for $\lambda \neq -1, -2, \ldots$. To extend Hadamard's finite part to negative integer values of λ, we have to consider some modifications. We can define $\mathrm{Pf}\cdot\, x^{-l}$ for $l = 1, 2, \ldots$ by adding $\mathrm{Pf}\cdot\, x^{-l}\theta(x)$ and $\mathrm{Pf}\cdot\, x^{-l}\theta(-x)$ formally, because then the troublemaking term $\varepsilon^0/0$ in (A-9) cancels out. It satisfies

$$(d/dx)\,\mathrm{Pf}\cdot\, x^{-l} = (-l)\,\mathrm{Pf}\cdot\, x^{-l-1} \tag{A-11}$$

We note that $\mathrm{Pf}\cdot\, x^{-l}$ is a generalization of Cauchy's principal value because

$$\mathrm{Pf}\cdot\, x^{-1} = \mathrm{P}\cdot\,(1/x) \tag{A-12}$$

Another way of avoiding difficulty at $\lambda = -l$ is to consider

$$\begin{aligned} Y_\nu(x) &= [\Gamma(\nu)]^{-1}\,\mathrm{Pf}\cdot\, x^{\nu-1}\,\theta(x) \quad \text{for} \quad \nu \neq 0, -1, -2, \ldots \\ &= \delta^{(m)}(x) \quad \text{for} \quad \nu = -m = 0, -1, -2, \ldots \end{aligned} \tag{A-13}$$

Then as seen from the definition (A-9), the integral

$$\int Y_\nu(x)\,\varphi(x)\,dx \tag{A-14}$$

becomes an entire function of ν.

Distributions of the following type are also important:

$$\lim_{\varepsilon\to+0} 1/(x \pm i\varepsilon)^\lambda \tag{A-15}$$

where the limit should be taken after the integration over x is carried out. The distributions (A-15) are related to Hadamard's finite part as follows:

$$\lim_{\varepsilon\to+0} 1/(x \pm i\varepsilon)^l = \mathrm{Pf}\cdot x^{-l} \mp i\pi[(l-1)!]^{-1}\,\delta^{(l-1)}(-x) \quad \text{for} \quad l = 1, 2, \ldots \tag{A-16}$$

$$\operatorname{Im} \lim_{\varepsilon\to+0} 1/(x \pm i\varepsilon)^\lambda = \mp\pi[\Gamma(\lambda)]^{-1}\,Y_{-\lambda+1}(-x) \tag{A-17}$$

If a test function $f(x)$ is an analytic function, then the integral

$$\int dx\, f(x)/(x \pm i\varepsilon)^l \tag{A-18}$$

can be regarded as a contour integral (see the next subsection) along real axis but avoiding the origin by using a small semicircle. In general, distributions can be defined also as boundary values of analytic functions [5].

A-2 Analytic Functions

A function $f(z)$ of a complex variable z is *holomorphic* at $z = a$ if the derivative df/dz at $z = a$ exists independently of the direction of the increment dz. If $f(z)$ is holomorphic at every point in a domain D, $f(z)$ is said to be holomorphic in D. Let Γ be a (Jordan) curve entirely lying in a domain D. Then one can define a *contour integral*

$$\int_\Gamma f(\zeta)\,d\zeta \tag{A-19}$$

in a way quite analogous to the definition of the curvilinear integral. If D is simply-connected, that is, if every closed curve lying in D can be continuously shrunk into a point of D, then (A-19) is dependent only on the two end points of Γ but not on the shape of Γ, that is, (A-19) vanishes if

Γ is closed (*Cauchy's theorem*). An important theorem for a holomorphic function is *Cauchy's formula*, which reads

$$f^{(m)}(z) = \frac{m!}{2\pi i} \int_{\Gamma} \frac{f(\zeta)}{(\zeta - z)^{m+1}} d\zeta \qquad (m = 0, 1, 2, \ldots) \qquad \text{(A-20)}$$

where Γ is any closed curve in D surrounding once the point $\zeta = z$ counterclockwise. From (A-20) together with the expansion of $(\zeta - z)^{-1}$ in powers of $z - a$ for $a \in D$, $f(z)$ is expanded into a *Taylor series*

$$f(z) = \sum_{m=0}^{\infty} f^{(m)}(a)\,(z - a)^m/m! \qquad \text{(A-21)}$$

in a disc $|z - a| < r$ entirely lying in D.

If two functions $f(z)$ and $g(z)$ holomorphic in D coincide with each other in a subset of D containing an accumulating point, then we have $f(z) \equiv g(z)$ everywhere in D. This property is known as the *theorem of unicity*. If there is a function $\tilde{f}(z)$ holomorphic in $\tilde{D} \supset D$ (but $\tilde{D} \neq D$) which coincides with $f(z)$ in D, $\tilde{f}(z)$ is called the *analytic continuation* of $f(z)$ to $\tilde{D}$. As seen above, the analytic continuation is unique. Let $f_j(z)$ be holomorphic in $D_j (j = 1, 2)$ and suppose $f_1(z) = f_2(z)$ in a connected subdomain D_0 of $D_1 \cap D_2$. Then we can define the analytic continuation $f(z)$ as above. If $D_1 \cap D_2$ is disconnected, however, we may not have $f_1(z) = f_2(z)$ in $D_1 \cap D_2 - D_0$, that is, $f(z)$ may not be single-valued in $D_1 \cup D_2$. To make it single-valued, one usually introduces *Riemann sheets*. When ew analytically continue $f(z)$ further, Riemann sheets may become more complicated, but the number of sheets at any point z is at most countable. The function $f(z)$, defined on each Riemann sheet, is globally called an *analytic function* of z.

A *singularity* of $f(z)$ is a point on a Riemann sheet to which it cannot be analytically continued, that is, $f(z)$ is not holomorphic at a singularity. If $f(z)$ is singular at $z = a$ but not in a neighborhood $N(a)$, the point $z = a$ is called an *isolated singularity*. When we go around $z = a$ once counterclockwise in $N(a)$ and return to the same point, the value of $f(z)$ may in general be different from the original one. This difference is called the *discontinuity*. If it is not equal to zero, the singularity $z = a$ is called a *branch point*. Otherwise, we can expand $f(z)$ into a *Laurent series*:

$$f(z) = \sum_{m=-\infty}^{+\infty} \alpha_m (z - a)^m \qquad \text{(A-22)}$$

in $0 < |z - a| < r$ in which $f(z)$ is holomorphic. Since $z = a$ is a singularity, at least one of α_m for $m < 0$ has to be nonzero. If $\alpha_m = 0$ for all $m < -k < 0$

but $\alpha_{-k} \neq 0$, then the point $z = a$ is called a *pole* of order k. For $k = 1$, it is called a *simple pole*; otherwise it is a *multiple pole*. If there exists no such k, the point is called an *essential singularity*. Near an essential singularity, $f(z)$ approaches to any value.

It is customary to regard the infinity as a point $z = \infty$; the behavior of $f(z)$ near $z = \infty$ is defined by that of $f(1/\zeta)$ near $\zeta = 0$. The function which is holomorphic everywhere (including $z = \infty$) is a constant alone. The functions which are holomorphic except at $z = \infty$ are called *entire*. Entire functions other than polynomials have an essential singularity at $z = \infty$. The functions which have no singularities other than poles in D are said to be *meromorphic* in D.

An analytic function may in general have nonisolated singularities. If it is singular along a closed curve, one cannot analytically continue the function further. Then this curve is called a *natural boundary*. Given any closed curve, there exists a function which has it as a natural boundary.

A function $f(z)$ of several complex variables† $z = (z_1, \ldots, z_n)$ is *holomorphic* at $z = a \equiv (a_1, \ldots, a_n)$ if it is totally differentiable. According to Hartogs, this definition is equivalent to an apparently weaker statement that $f(z)$ has all partial derivatives $\partial f/\partial z_j$ $(j = 1, \ldots, n)$. If and only if $f(z)$ is holomorphic at $z = a$, it can be expanded into a multiple Taylor series around $z = a$.

Most properties of the function of one variable such as Cauchy's formula can be extended into the case of the function of several variables. The important difference between them is the structure of singularities. Any function of more than one variable cannot have an isolated singularity. More precisely, any analytic function cannot have a singularity of $2n - 3$ or less dimensions which is disconnected from other singularities. Contrary to the function of one variable, therefore, given a domain D, there does not necessarily exist a function which cannot be continued analytically beyond D. If such a function exists, D is called a *natural domain of holomorphy*. Its geometrical shape is characterized by a word "pseudoconvex" (Oka's theorem), but we do not give its definition here. If D is not a natural domain of holomorphy, every function holomorphic in D is automatically continued to a certain natural domain of holomorphy including D. The latter is called the *envelope of holomorphy* of D and to find it is called the *analytic completion*.

Finally, we discuss some analytic properties of integrals involving analytic

† There are some textbooks on the function of several complex variables [6, 7].

parameters. Let $f(z, x)$ be holomorphic in both z and x, and consider a function defined by a contour integral

$$F(z) = \int_{\Gamma} f(z, x)\, dx \tag{A-23}$$

where the contour Γ is a curve lying in a domain in which $f(z, x)$ is holomorphic in x. The function $F(z)$ is holomorphic in a domain in which (A-23) is well defined. Now, we analytically continue $f(z, x)$ in z; then we shall meet some singularities on Γ because their locations in the x plane are, in general, functions of z. Suppose that they are all isolated. If only one singularity comes across Γ, $F(z)$ will not have a singularity at such a value of z, because we can deform the contour Γ by Cauchy's theorem in order to avoid the singularity. The contour deformation becomes impossible *either* if a singularity comes to an end point of Γ *or* if two singularities approach to Γ from the opposite sides. Then we have a singularity of $F(z)$. The former case is called an *end-point singularity*, while the latter situation is a *pinch* [8].

When x is a set of n variables (z may also represent several variables) and (A-23) is regarded as a multiple contour integral, we can make similar considerations. The contour Γ is now a region of n dimensions. A singularity of $F(z)$ appears when x belongs to an m-dimensional face $\Gamma^{(m)}$ of Γ $(0 \leqq m \leqq n)$ and a pinch occurs for the remaining $n - m$ dimensions. Here a pinch is a situation such that several singularity hypersurfaces approach to Γ in such a way that Γ cannot be deformed to avoid the singularities. A more precise description of the singularities of $F(z)$ can be made by the homological method [9], in which a pinch corresponds to the existence of a "vanishing cycle". If the position of a pinch in x is isolated, then it is called a *simple pinch.*

When $f(z, x)$ is not an analytic function of x but a distribution of $x = (x_1, \ldots, x_n)$ and Γ is a certain integration region in an n-dimensional euclidean space, (A-23) is called an *integral representation* of $F(z)$. Usually, $f(z, x)$ is a product of an elementary analytic function, $g(z, x)$, of both z and x such as a rational function and a distribution, $\varphi(x)$, of x alone. Then we can see that $F(z)$ is holomorphic at every point $z = a$ such that the singularity region of $g(a, x)$ and the support of $\varphi(x)$ are disjoint. Therefore it is important to find the support of $\varphi(x)$; this distribution is called a *weight distribution* of the integral representation.

In particular, if a function $F(z)$ of a single variable z is holomorphic except for the positive real axis and if $F(z) = 0(z^{-\varepsilon})$ $(\varepsilon > 0)$ as $z \to \infty$, then $F(z)$ has the following integral representation:

$$F(z) = \int_{-0}^{\infty} dx \frac{\varphi(x)}{x - z} \tag{A-24}$$

which is called a (single) *spectral representation* of $F(z)$. If $z = (z_1, \ldots, z_n)$, we can likewise define a multiple spectral representation.

References

(1) L. Schwartz, *Théorie des Distributions* (Hermann & Cie, Paris, 1950).

(2) Ref. 1, Chap. 7.

(3) Ref. 1, Chap. 3.

(4) Ref. 1, Chap. 2.

(5) M. Sato, *J. Faculty of Science, Univ. of Tokyo* **8**, 139 (1959); **8**, 387 (1960).

(6) S. Bochner and W. T. Martin, *Several Complex Variables* (Princeton University Press, Princeton, N. J., 1948).

(7) V. S. Vladimirov, *Method of Theory of Several Complex Variables* (MIT Press, Cambridge, Mass., 1966).

(8) R. J. Eden, P. V. Landshoff, D. I. Olive, and J. C. Polkinghorne, *The Analytic S-Matrix* (Cambridge University Press, London, 1966).

(9) R. C. Hwa and V. L. Teplitz, *Homology and Feynman Integrals* (W. A. Benjamin, Inc., N. Y., 1966).

Appendix B Feynman Integrals for Single-Loop Graphs

B-1 General Remarks and Self-Energy

As the simplest, nontrivial examples, the Feynman integrals corresponding to the Feynman graphs consisting of only one circuit are of special interest. In this appendix, we summarize some detailed results concerning them without derivation.

Without loss of generality, we may confine ourselves to (strongly connected) single-loop graphs G which have no internal vertices. Then we have $N = M = n$. It is convenient to denote internal lines of G by $1, 2, \ldots, n$ and external vertices by $a_{12}, a_{23}, \ldots, a_{n-1,n}, a_{n,n+1} \equiv a_{n1}$ in such a way that each line j is incident with $a_{j-1,j}$ and $a_{j,j+1}$ (see Fig. B-1).

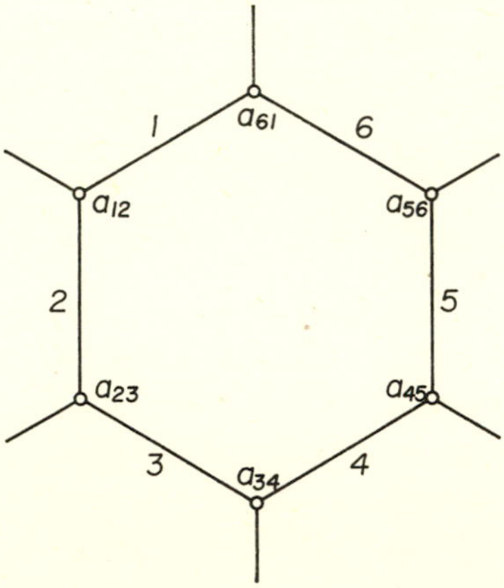

Fig. B-1 A single-loop Feynman graph.

According to the modified cut-set representation (7-46) of V, we have

$$UV = \tfrac{1}{2} \sum_{i,j=1}^{n} \alpha_i \alpha_j (m_i^2 + m_j^2 - p_{ij}^2) \tag{B-1}$$

with $U = \sum_j \alpha_j$, where

$$p_{ij} \equiv \sum_{k=i}^{j-1} p_{k,k+1} \quad \text{for} \quad j > i$$
$$p_{ii} \equiv 0, \quad p_{ji} = -p_{ij} \tag{B-2}$$

and $p_{k,k+1}$ is the external momentum outgoing from $a_{k,k+1}$. The overall momentum conservation reads

$$\sum_{i=1}^{n} p_{i,i+1} = p_{1,n+1} \equiv p_{11} = 0 \tag{B-3}$$

Let

$$s_{ij} \equiv p_{ij}^2, \quad (s_{ii} \equiv 0)$$
$$v_{ij} \equiv s_{ij} - m_i^2 - m_j^2, \quad (v_{ii} = -2m_i^2) \tag{B-4}$$
$$\sigma_{ij} \equiv -v_{ij}/2m_i m_j, \quad (\sigma_{ii} = 1)$$

The stability conditions for an external mass $s_{ij}^{\frac{1}{2}}$ and for internal masses m_i and m_j are written as

$$\sigma_{i,i+1} \geqq -1$$
$$\sigma_{i,i+1} \leqq +1 \tag{B-5}$$

respectively. For $|\sigma_{i,i+1}| \leqq 1$ ($i = 1, \ldots, n$), we set

$$\sigma_{i,i+1} \equiv \cos \theta_{i,i+1} \quad (0 \leqq \theta_{i,i+1} \leqq \pi) \tag{B-6}$$

The location of the leading singularity is determined by

$$\det (v_{ij} \mid i, j = 1, \ldots, n) = 0 \tag{B-7}$$

When we solve (B-7) with respect to a particular invariant square s_{ij}, we have two solutions. If they are real, the larger one gives the location of the leading anomalous threshold when it is present.

For $n \geqq 3$, the Feynman integral corresponding to G is

$$f_G(s) = (n-3)! \int_{\Delta} d^{n-1} \alpha (UV - i\varepsilon)^{-n+2} \tag{B-8}$$

For $n = 2$, the self-energy Feynman integral is ultraviolet-divergent. Its explicit formula is

$$f_G(s) = \frac{R^{\frac{1}{2}}}{2s} \log \frac{-v + R^{\frac{1}{2}}}{-v - R^{\frac{1}{2}}} - \frac{m_1^2 - m_2^2}{2s} \log \frac{m_1^2}{m_2^2} + \text{const} \tag{B-9}$$

with $s \equiv s_{12}$, $v \equiv v_{12}$, and $R \equiv \lambda(m_1^2, m_2^2, s)$, where

$$\lambda(x, y, z) \equiv x^2 + y^2 + z^2 - 2(xy + yz + zx) \tag{B-10}$$

The constant in (B-9) is infinite, and should be made finite by renormalization. The imaginary part is free from divergence:

$$\pi^{-1} \operatorname{Im} f_G(s) = (R^{\frac{1}{2}}/s)\, \theta(s - (m_1 + m_2)^2) \tag{B-11}$$

B-2 Vertex Function

For $n = 3$, it is convenient to write s_{ij} as s_k and v_{ij} as v_k, where (i, j, k) is a permutation of (1, 2, 3). We define the following functions:

$$\begin{aligned}
\Phi &\equiv \det(v_{ij} \mid i, j = 1, 2, 3) \\
&= 2\Bigg[\sum_{k=1}^{3} m_k^4 s_k + \sum_{1 \leqq j < k \leqq 3} m_j^2 m_k^2 (s_i - s_j - s_k) \\
&\qquad + \sum_{k=1}^{3} m_k^2 s_k (s_k - s_i - s_j) + s_1 s_2 s_3\Bigg] \\
L_k &\equiv \tfrac{1}{2}\, \partial\Phi/\partial m_k^2 \\
&= s_k^2 - s_k(s_i + s_j + m_i^2 + m_j^2 - 2m_k^2) - (m_i^2 - m_j^2)(s_i - s_j) \\
R_k &\equiv -\begin{vmatrix} v_{ii} & v_{ij} \\ v_{ji} & v_{jj} \end{vmatrix} = \lambda(m_i^2, m_j^2, s_k) \\
\Lambda &\equiv \lambda(s_1, s_2, s_3)
\end{aligned} \tag{B-12}$$

There are four identities:

$$\begin{aligned}
L_k^2 - \Lambda R_k &= 2 s_k \Phi \qquad (k = 1, 2, 3) \\
\sum_{k=1}^{3} L_k &= \Lambda
\end{aligned} \tag{B-13}$$

The explicit formula takes simple form if we consider

$$g_G(s) \equiv -\left(\sum_{k=1}^{3} \partial/\partial m_k^2\right) f_G(s) \tag{B-14}$$

rather than $f_G(s)$. We have [1]

$$g_G = \Phi^{-1} \sum_{k=1}^{3} L_k g_{G/k} \tag{B-15}$$

where

$$g_{G/k} \equiv \int_0^1 \frac{d\alpha}{\alpha m_i^2 + (1-\alpha) m_j^2 - \alpha(1-\alpha) s_k}$$

$$= \frac{1}{R_k^{\frac{1}{2}}} \log \frac{-v_k + R_k^{\frac{1}{2}}}{-v_k - R_k^{\frac{1}{2}}} \tag{B-16}$$

with

$$\pi^{-1} \operatorname{Im} g_{G/k} = 2R_k^{-\frac{1}{2}} \theta(s_k - (m_i + m_j)^2) \tag{B-17}$$

The formula (B-15) may be rewritten as

$$\begin{vmatrix} g_G & -g_{G/1} & -g_{G/2} & -g_{G/3} \\ 1 & v_{11} & v_{12} & v_{13} \\ 1 & v_{21} & v_{22} & v_{23} \\ 1 & v_{31} & v_{32} & v_{33} \end{vmatrix} = 0 \tag{B-18}$$

The imaginary part of the Feynman integral with respect to s_3 is as follows [2, 3, 4]. Let $|\sigma_{23}| \leqq 1$ and $|\sigma_{31}| \leqq 1$. The contribution from the normal threshold located at $s_3 = (m_1 + m_2)^2$ is

$$\pi^{-1}[\operatorname{Im} f_G(s)]_{\text{normal}} = \varrho_3(s)\, \theta(s_3 - (m_1 + m_2)^2) \tag{B-19}$$

where

$$\begin{aligned} \varrho_3(s) &= \frac{1}{\Lambda^{\frac{1}{2}}} \log \frac{L_3 + (\Lambda R_3)^{\frac{1}{2}}}{L_3 - (\Lambda R_3)^{\frac{1}{2}}} && \text{for } \Lambda > 0 \\ &= 2R_3^{\frac{1}{2}}/L_3 && \text{for } \Lambda = 0 \\ &= \frac{2}{(-\Lambda)^{\frac{1}{2}}} \tan^{-1} \frac{(-\Lambda R_3)^{\frac{1}{2}}}{L_3} && \text{for } \Lambda < 0 \end{aligned} \tag{B-20}$$

The condition $\Lambda \geqq 0$ is equivalent to $s_3 \geqq (s_1^{\frac{1}{2}} + s_2^{\frac{1}{2}})^2$, namely, the condition for the s_3 physical region, in which the Cutkosky rule for the discontinuity formula (see Subsection 14-2) is valid; we can also show that $L_3 > 0$ there. If

$$m_1 + m_2 < s_1^{\frac{1}{2}} + s_2^{\frac{1}{2}} \tag{B-21}$$

we have to take account of the $\Lambda < 0$ case of (B-20) (the analytic continuation of the $\Lambda > 0$ case), which exhibits the correct normal threshold behavior at $s_3 = (m_1 + m_2)^2$ if and only if $L_3 > 0$ at $s_3 = (m_1 + m_2)^2$; in this case, $\tan^{-1}$ takes its principal value.

If $L_3 < 0$ at $s_3 = (m_1 + m_2)^2$, we have an anomalous threshold $s_3 = \hat{s}_3$ in $(m_1 - m_2)^2 < s_3 < (m_1 + m_2)^2$, that is, it appears if and only if

$$\sigma_{23} + \sigma_{31} < 0, \quad \text{i.e.,} \quad \theta_{23} + \theta_{31} > \pi \tag{B-22}$$

Its location $\hat{s}_3$ is the larger root of $\Phi = 0$, i.e.,

$$\theta_{12} + \theta_{23} + \theta_{31} = 2\pi \tag{B-23}$$

From (B-22) together with $|\sigma_{23}| \leqq 1$ and $|\sigma_{31}| \leqq 1$, it is easy to show that (B-21) is always satisfied. Since $L_3 > 0$ at $s_3 = (s_1^{\frac{1}{2}} + s_2^{\frac{1}{2}})^2$ and $L_3 < 0$ at $s_3 = (m_1 + m_2)^2$, L_3 vanishes once between those two points. Hence $\tan^{-1}$ in (B-20) passes through $\pi/2$ and it becomes π at $s_3 = (m_1 + m_2)^2$. The imaginary part of $f_G(s)$ for $s_3 < (m_1 + m_2)^2$ due to the anomalous threshold is

$$\pi^{-1}[\operatorname{Im} f_G(s)]_{\text{anomalous}} = 2\pi(-\Lambda)^{-\frac{1}{2}}\, \theta(s_3 - \hat{s}_3)\, \theta[(m_1 + m_2)^2 - s_3] \tag{B-24}$$

so that the total imaginary part

$$\operatorname{Im} f_G(s) = [\operatorname{Im} f_G(s)]_{\text{normal}} + [\operatorname{Im} f_G(s)]_{\text{anomalous}} \tag{B-25}$$

is still continuous at $s_3 = (m_1 + m_2)^2$. We note that $\operatorname{Im} f_G(s) > 0$ for $s_3 > \hat{s}_3$.

If $s_3 \leqq (m_1 - m_2)^2$, then the vertex function has a double spectral representation [4, 5, 6, 7,]:

$$f_G(s) = \int_0^\infty ds_1' \int_0^\infty ds_2' \frac{\sigma_{12}(s_1', s_2', s_3)}{(s_1' - s_1)(s_2' - s_2)} \tag{B-26}$$

where

$$\sigma_{12}(s_1, s_2, s_3) \equiv \Lambda^{-\frac{1}{2}}\theta(-\Phi)\, \theta(s_1 - (m_2 + m_3)^2)\, \theta(s_2 - (m_3 + m_1)^2)$$
$$\text{for} \quad s_3 \leqq 0 \qquad \text{(B-27)†}$$

and there is an additional term due to the second-type singularity for $s_3 > 0$. For $(m_1 - m_2)^2 < s_3 < (m_1 + m_2)^2$, we have complex singularities in (s_1, s_2) space [8, 3].

Nambu's triple spectral representation is valid only when $m_1 = m_2 = m_3 = 0$. The weight distribution then is given by [8]

$$\pi^{-1}(-\Lambda)^{-\frac{1}{2}}\, \theta(-\Lambda)\, \theta(s_1)\, \theta(s_2)\, \theta(s_3) \tag{B-28}$$

† One may omit either of the last two θ functions, owing to $\theta(-\Phi)$.

The two-variable PTIR [cf. (19-20)] is as follows [8]:

$$f_G(s) = \int_0^1 dz \int_{-\infty}^{\infty} d\gamma \frac{\varphi_{12}(z, \gamma)}{\gamma - zs_1 - (1 - z)\, s_2} \tag{B-29}$$

where

$$\varphi_{12}(z, \gamma) \equiv \frac{\beta - \eta + m_3^2}{\beta D^{\frac{1}{2}}} \theta(\beta - (\eta^{\frac{1}{2}} + m_3)^2)\, \theta(\eta) + \frac{\beta - \eta + m_3^2 - D^{\frac{1}{2}}}{2\beta D^{\frac{1}{2}}} \theta(-\eta) \tag{B-30}$$

with

$$\begin{aligned} \beta &\equiv \gamma - z(1 - z)\, s_3 \\ \eta &\equiv (1 - z)\, m_1^2 + zm_2^2 - z(1 - z)\, s_3 \\ D &\equiv \lambda(\beta, \eta, m_3^2) \end{aligned} \tag{B-31}$$

If $s_3 \leqq (m_1 + m_2)^2$ then $\eta \geqq 0$, whence the first term alone survives in (B-30). If $s_3 > (m_1 + m_2)^2$ then the support of φ_{12} is not bounded below in γ.

B-3 Scattering Amplitude

Throughout this subsection, we assume $|\sigma_{i,i+1}| \leqq 1$ $(i = 1, 2, 3, 4)$, and write $s_{13} \equiv s$ and $s_{24} \equiv t$.

The conditions for the physical region in which the scattering in the s channel can take place are

$$\begin{gathered} s \geqq \max\,[(s_{12}^{\frac{1}{2}} + s_{23}^{\frac{1}{2}})^2, (s_{34}^{\frac{1}{2}} + s_{41}^{\frac{1}{2}})^2] \\ |\cos\theta| \leqq 1 \end{gathered} \tag{B-32}$$

where θ stands for the scattering angle in the center-of-mass system; more explicitly

$$\cos\theta = L_s/(\Lambda_2 \Lambda_4)^{\frac{1}{2}} \tag{B-33}$$

where

$$\begin{aligned} L_s &\equiv s^2 - s(s_{12} + s_{23} + s_{34} + s_{41} - 2t) - (s_{12} - s_{23})\,(s_{34} - s_{41}) \\ \Lambda_2 &\equiv \lambda(s_{12}, s_{23}, s) \\ \Lambda_4 &\equiv \lambda(s_{34}, s_{41}, s) \end{aligned} \tag{B-34}$$

The explicit formula for $f_G(s, t)$ is obtainable but complicated [9].

The conditions for the presence of anomalous thresholds are as follows [10]:

(a) Leading

$$2\pi < \theta_{12} + \theta_{23} + \theta_{34} + \theta_{41} < 2\pi + 2 \min(\theta_{12}, \theta_{23}, \theta_{34}, \theta_{41}) \quad \text{(B-35)}$$

(b) Nonleading

$$\begin{aligned} &(s \text{ channel}) \quad \theta_{12} + \theta_{23} > \pi \quad \text{or} \quad \theta_{34} + \theta_{41} > \pi \\ &(t \text{ channel}) \quad \theta_{23} + \theta_{34} > \pi \quad \text{or} \quad \theta_{41} + \theta_{12} > \pi \end{aligned} \quad \text{(B-36)}$$

The imaginary part due to the normal threshold in the s channel is [11]†

$$\pi^{-1}[\operatorname{Im} f_G(s, t)]_{\text{normal}} = \frac{1}{\Psi^{\frac{1}{2}}} \log \frac{-\Phi_{24} + (R_{13}\Psi)^{\frac{1}{2}}}{-\Phi_{24} - (R_{13}\Psi)^{\frac{1}{2}}} \theta(s - (m_1 + m_3)^2) \quad \text{(B-37)}$$

where

$$\Psi \equiv \det(v_{ij} \mid i, j = 1, 2, 3, 4)$$

$$\Phi_{24} = \begin{vmatrix} v_{11} & v_{12} & v_{13} \\ v_{31} & v_{32} & v_{33} \\ v_{41} & v_{42} & v_{43} \end{vmatrix} = \tfrac{1}{2}\, \partial\Psi/\partial t$$

$$R_{13} \equiv \lambda(m_1^2, m_3^2, s) = \tfrac{1}{2}\, \partial^2\Psi/\partial t^2 \quad \text{(B-38)}$$

Because of Jacobi's theorem, we have an identity

$$\Phi_{24}^2 - R_{13}\Psi = \Phi_{22}\Phi_{44} \quad \text{(B-39)}$$

where Φ_{ii} is the (i, i) cofactor of Ψ, namely, the Φ of G/i, and $\Phi_{ii} > 0$ $(i = 2, 4)$ for $s > (m_1 + m_3)^2$.

The imaginary part due to an anomalous threshold can be obtained by deforming the branch cut of the normal threshold according to the movement of the anomalous threshold [12, 13].

The Mandelstam representation

$$f_G(s, t) = \int_0^\infty ds' \int_0^\infty dt' \, \frac{\sigma_{st}(s', t')}{(s' - s)(t' - t)} \quad \text{(B-40)}$$

† For $\Psi \leqq 0$, we should analytically continue the right-hand side as in (B-20).

is possible if and only if [14]

$$\theta_{12} + \theta_{23} + \theta_{34} + \theta_{41} \leqq 2\pi \tag{B-41}$$

If there is no anomalous threshold, the double spectral function is given by [11, 15]

$$\sigma_{st}(s, t) = 2\Psi^{-\frac{1}{2}}\theta(\Psi)\,\theta(s - (m_1 + m_3)^2)\,\theta(t - (m_2 + m_4)^2) \qquad \text{(B-42)}\dagger$$

B-4 Production Amplitudes

For $n = 5$, the leading singularity is a simple pole [16]. The explicit formula for f_G can be written as [17, 18]‡

$$\begin{vmatrix} f_G & -f_{G/1} & \cdots & -f_{G/5} \\ 1 & v_{11} & \cdots & v_{15} \\ 1 & v_{21} & \cdots & v_{25} \\ \cdot & \cdot & \cdot & \cdot \\ 1 & v_{51} & \cdots & v_{55} \end{vmatrix} = 0 \tag{B-43}$$

where $f_{G/j}$ is a Feynman integral for $n = 4$. The analyticity of f_G is unfavorable to multiple spectral representations [19, 20].

For $n \geqq 6$, there is no leading singularity [21, 22]. If we replace any row of a matrix $(v_{ij} \mid i, j = 0, 1, \ldots, n)$ with $v_{00} \equiv 0$ and $v_{0i} \equiv v_{i0} \equiv 1$ $(i = 1, \ldots, n)$ by $(f_G, -f_{G/1}, \ldots, -f_{G/n})$, then the resultant matrix is of rank 6 [18, 23, 24].

References

(1) G. Källén, and A. Wightman, *Mat. Fys. Skr. Dan. Vid. Selsk*, **1**, No. 6 (1958).
(2) K. Yamamoto, *Progr. Theoret. Phys.* **25**, 361 (1961).
(3) P. V. Landshoff, and S. B. Treiman, *Phys. Rev.* **127**, 649 (1962).
(4) C. Fronsdal and R. E. Norton, *J. Math. Phys.* **5**, 100 (1964).
(5) K. Yamamoto, *Progr. Theoret. Phys.* **25**, 720 (1961).
(6) C. Fronsdal, K. T. Mahanthappa, and R. E. Norton, *Phys. Rev.* **127**, 1847 (1962).
(7) Yu. A. Simonov, *J. E. T. P.* **43**, 2263 (1962), [translation: *Sov. Phys. J. E. T. P.* **16**, 1599 (1963).
(8) N. Nakanishi, *Progr. Theoret. Phys.* **24**, 1275 (1960).
(9) A. C. T. Wu, *Mat. Fys. Medd. Dan. Vid. Selsk.* **33**, No. 3 (1961).

† One may omit either of the last two θ functions, owing to $\theta(\Psi)$.
‡ In Ref. [17], a numerical factor is erroneous.

(10) R. Karplus, C. M. Sommerfield, and E. H. Wichmann, *Phys. Rev.* **114**, 376 (1959).
(11) S. Mandelsatm, *Phys. Rev.* **115**, 1741 (1959).
(12) S. Mandelstam, *Phys. Rev. Letters* **4**, 84 (1960).
(13) J. Otokozawa, *Progr. Theoret. Phys.* **25**, 277 (1961).
(14) C. Fronsdal, R. E. Norton, and K. T. Mahanthappa, *J. Math. Phys.* **4**, 859 (1963).
(15) T. W. B. Kibble, *Phys. Rev.* **117**, 1159 (1960).
(16) R. E. Cutkosky, *J. Math. Phys.* **1**, 429 (1960).
(17) F. R. Halpern, *Phys. Rev. Letters* **10**, 310 (1963).
(18) D. B. Melrose, *Nuovo Cim.* **40**, 181 (1965).
(19) L. F. Cook, Jr. and J. Tarski, *J. Math. Phys.* **3**, 1 (1962).
(20) O. I. Zav'yalov and V. P. Pavlov, *J. E. T. P.* **44**, 1500 (1963), [translation: *Sov. Phys. J. E. T. P.* **17**, 1009 (1963)].
(21) L. M. Brown, *Nuovo Cim.* **22**, 178 (1961).
(22) V. E. Asribekov, *J. E. T. P.* **43**, 1826 (1962), [translation: Sov. Phys. *J. E. T. P.* **16**, 1289 (1963)].
(23) G. Källén, and J. S. Toll, *J. Math. Phys.* **6**, 299 (1965).
(24) B. Petersson, *J. Math. Phys.* **6**, 1955 (1965).

Author Index

(f indicates a footnote, and * stands for a page of References.)

Subject Index

(f indicates a footnote.)